KOREAN MULTINATIONALS IN EUROPE

KOREAN MULTINATIONALS IN EUROPE

Judith Cherry

CURZON

First Published in 2001
by Curzon Press
Richmond, Surrey
http://www.curzonpress.co.uk

Typeset in Stempel Garamond by LaserScript Ltd, Mitcham, Surrey
Printed and bound in Great Britain by
TJ International, Padstow, Cornwall

British Library Cataloguing in Publication Data
A catalogue record of this book is available from the British Library

Library of Congress Cataloguing in Publication Data
A catalogue record for this book has been requested

ISBN 0-7007-1480-4

Contents

List of tables vii
List of figures ix
Acknowledgements x
List of Abbreviations xi
Romanization and Reference System xiii

Introduction 1

1 A Historical Overview of the Development of the Theory of Foreign Direct Investment and the Multinational Enterprise 4
Early Theories of Foreign Production 4
The 1960s and 1970s 5
The late 1970s and 1980s 8
Motivations for Investment by MNEs from Industrialized Countries 11
Theories of Developing Country FDI 12
Motivations for Investment by MNEs from Developing Countries 15
Conclusions 16

2 Foreign Direct Investment Theory: The Korean Perspective 18
A Chronological Survey of the Korean Literature 19
Korean Theories of Foreign Direct Investment 25
Motivations for Investment 33
Conclusions 33

3 The Economic Context: The Republic of Korea (1948–1997) 36
The Growth and Development of the Korean Economy 36
The *Chaebŏl* 49
Conclusions 55

4 The Development and Characteristics of the Korean Consumer Electronics Industry 57
Korea's Place in the Global Electronics Industry 58
A Historical Overview of the Development of the Korean Consumer Electronics Industry 61
The Characteristics of the Korean Consumer Electronics Industry 71
Conclusions 80

5 Korean Foreign Direct Investment 82
A Chronological Overview 83
Korean Manufacturing Investment 96
Investment in Industrialized countries and Developing Countries 101
Conclusions 104

6 Korea's Relations with Europe: An Overview 107
Korea and Europe: A Historical Perspective 107
The Asia-Europe Meeting (ASEM) 112
ROK-EU Economic Relations 114
South Korea's Relations with Eastern Europe 126
Conclusions 128

7 Korean Foreign Direct Investment in Europe 130
Overview 130
Korean Consumer Electronics Investment 142
Conclusions 148

8 Korean Consumer Electronics Investment in Europe: A Theoretical/Macroeconomic Perspective 151
Korean Consumer Electronics Investment in Europe: Playing 'Follow-the-Leader' or 'Leapfrog'? 154
Korean Consumer Electronics Investment in Europe: A Theoretical/Macroeconomic Perspective 160

Summary and Conclusions 171

Postscript: The *Chaebŏl* and the Korean Economy in the 'IMF era' 176

Notes 193
Appendices 197
Bibliography 206
Index 235

Tables

3.1 The Top 10 *Chaebŏl*, 1996. 49

4.1 Global Electronics Production, 1995–1996. 59
4.2 Global Electronics Exports, 1995. 60
4.3 Hourly Manufacturing Wage Rates, 1980–1990. 67
4.4 The Korean Electronics Industry, 1988–1993. 69
4.5 The Big Three's Share of Consumer Electronics Production by Item, 1988. 78

5.1 Korean Foreign Direct Investment, 1968–1996. 83
5.2 Korean Foreign Direct Investment: By Sector, 1968–1979. 85
5.3 Korean Foreign Direct Investment: By Region, 1968–1979. 86
5.4 Korean Foreign Direct Investment: By Sector, 1980–1985. 87
5.5 Korean Foreign Direct Investment: By Region, 1980–1985. 88
5.6 Korean Foreign Direct Investment: By Sector, 1986–1996. 92
5.7 Korean Foreign Direct Investment: By Region, 1986–1996 93
5.8 Korean Foreign Direct Investment: By Sector, 1968–1996. 96
5.9 Korean Manufacturing Investment, 1968–1996. 96
5.10 Korean Manufacturing Investment: By Sector and Region, December 1996. 97
5.11 Korean Manufacturing Investment: By Top Five Host Countries, December 1996. 98
5.12 Korean Manufacturing Investment: By Sector, December 1996. 99
5.13 Korean Manufacturing Investment: By Sector and Region, December 1996. 100
5.14 Korean Foreign Direct Investment Projects: By Large Enterprises and SMEs, December 1996. 102
5.15 Korean Foreign Direct Investment: By Sector and AC/DC Markets, December 1996. 103

5.16 Korean Manufacturing Investment: By Sector and AC/DC Markets, December 1996. 103
5.17 Korean Exports and FDI: Leading Sectors, 1990 & 1994. 104

6.1 ROK-EU Trade, 1990–1996. 117
6.2 Investment in Korea by Major Western European Nations, 1962–1996. 119
6.3 Trends in Bilateral Trade, South Korea and Eastern Europe, 1990–1996. 128

7.1 Korean Foreign Direct Investment in Europe and the World: Annual Growth Rates, 1968–1996. 132
7.2 Korean Foreign Direct Investment: By Region, December 1996. 132
7.3 Korean Investment in Europe; By Country, December 1996. 133
7.4 Korean Investment in Europe: By Sector, December 1996. 134
7.5 Korean Manufacturing Investment: By Country, December 1996. 135
7.6 Korean Manufacturing Investment in Europe: By Sector, December 1996. 136
7.7 Top Five FDI Sectors in Europe: Percentage of Investment From Korea in Each Sector, December 1996. 137
7.8 Korean Trading Investment by Country, December 1996. 138
7.9 Korean Investment in Western, Central and Eastern Europe: By Sector, December 1996. 138
7.10 Korean Manufacturing Investment in Western, Central and Eastern Europe: By Sector, December 1996. 139
7.11 Average Size of Korean Investments in Europe, North America and South-east Asia, December 1996. 140
7.12 Equity Participation Ratio in Europe, North America and South-east Asia, December 1996. 141
7.13 Korean Global Electronics FDI Projects: By Sector, June 1997. 144
7.14 FDI Projects by Korean Electronics Companies, June 1997. 144
7.15 FDI by Korean Electronics Companies: By Country, June 1997. 146

8.1 Investment in Europe by the 'Big Three' Consumer Electronics Companies, June 1997. 155
8.2 Oligopolistic Reaction – FDI in Europe by the 'Big Three', June 1997. 157

Figures

2.1 Wŏn Chong-gŭn's Theoretical Explanation of Reverse Direct Investment. 21
2.2 Jun Yongwook's Theory of Structural Market Defence. 27
2.3 Yun Tong-jin's Post-Product Life Cycle Model: The Seven Stages of Trade and Investment for Korean Consumer Electronics Firms. 29
2.4 Yun Tong-jin's Post-Product Life Cycle Model. 30

8.1 The Business Environment and ROK Government Policy, 1963–1986. 162
8.2 Changes in the Business Environment and ROK Government Policy, 1987–1996. 164

Acknowledgements

Thanks are due first and foremost to KPMG, whose generous financial assistance made it possible for me to write my thesis, with special thanks to Mr Nigel Platts for his support and faith in me.

I also owe a great debt to the following people: Dr Ruth Grayson and Professor Steven Kirby for their invaluable comments and suggestions, Bill, Kath, Marjorie, Julia, Carol, Pete, and Rachel for keeping me sane while I wrote my thesis, Yongju and Jong Lee for their assistance and encouragement, and Kim Hae-Kun for his never-ending support and friendship.

My thanks also go to the staff of the Inter-Library Loans Section at Sheffield University Library, the National Library of Korea, the Electronics Industries Association of Korea, and the British Embassy, Seoul ... and also to all the friends in Seoul who have provided accommodation and support during my research trips to Korea.

Finally, a big 'Thank you' to all my students, past and present, for making my job the great joy that it is.

Abbreviations

AC	Advanced country
ACFDI	Advanced country foreign direct investment
APEC	Asia-Pacific Economic Cooperation
ASEM	Asia-Europe Meeting
ASEAN	Association of South-East Asian Nations
CEN	Comité Européen de Normalisation
CENELEC	Comité Européen de Normalisation Électrotechnique
CIS	Commonwealth of Independent States
DC	Developing country
DCFDI	Developing country foreign direct investment
DER	Debt-to-equity ratio
DPRK	Democratic People's Republic of Korea
DRAM	Dynamic random access memory
EC	European Community
ECDF	Economic Development Cooperation Fund
EFTA	European Free Trade Area
EIAK	Electronics Industries Association of Korea
EP	Equity participation
EU	European Union
FDI	Foreign direct investment
FOB	Free on board
FOR	Factory operating ratio
FSMA	Firm-specific monopolistic advantage
GATT	General Agreement on Trade and Tariffs
GDP	Gross domestic product
GNP	Gross national product
GSP	General System of Preferences
GTC	General trading company

HCIP	Heavy and Chemical Industry Plan
IDC	Investment Development Cycle
IMF	International Monetary Fund
IPR	Intellectual property rights
ITM	Integrated Theoretical Model
KFTA	Korea Foreign Trade Association
KIET	Korea Institute for Economics and Technology
KODECO	*Han'guk Nambang Kaebal Chushik Hoesa*
LDAC	Less-developed advanced country
LDC	Lesser developing country
M&A	Mergers and acquisitions
MNE	Multinational enterprise
NAFTA	North America Free Trade Association
NIE	Newly-Industrializing Economy
NOI	Net outward investment
OECD	Organization for Economic Cooperation and Development
OEM	Original equipment manufacture
OISC	Overseas Investment Screening Council
OLI	Ownership-location-internalization
OMA	Orderly Marketing Agreement
PLC	Product life cycle
PRC	People's Republic of China
R&D	Research and development
RDI	Reverse direct investment
ROK	Republic of Korea
SME	Small- and medium-sized enterprise
UK	United Kingdom
UN	United Nations
UNCTAD	United Nations Conference on Trade and Development
UPP	United People's Party
UR	Uruguay Round
US	United States
US$	US dollars
USA	United States of America
VCR	Video cassette recorder
VER	Voluntary Export Restrictions
WTO	World Trade Organization

Romanization and Reference System

The system of romanization used in this book is the McCune-Reischauer (M-R) system. Four of the Korean authors whose works appear in the bibliography have published in both English and Korean. In these cases, the author's romanization of his name is used throughout, even though this may conflict with the rules of the M-R system. The authors in question are Jun Yongwook, Choi Yong-rok, Kim Bak-soo, and Min Chung-ki. In addition, Jun Yongwook and Park Gi-han have both published in English under different names: in this book, Jun is referred to as Jun Yongwook (rather than Jun Young-wook) and Park is referred to as Park Gi-han (as opposed to Park Ki-an).

The Harvard reference system has been used throughout, with names given in full where authors share a surname and have published in the same year. This is most commonly the case with certain Korean names including Kim, Pak (Park) and Yi (Lee).

Introduction

In 1968, the Korea Development Corporation (KODECO) became the first Korean[1] company to engage in foreign direct investment (FDI), by establishing a joint venture lumber company in Indonesia. For almost two decades thereafter, Korean outward investment remained at low levels, with fewer than 70 new cases each year and less than US$160 million invested annually in projects overseas. However, between 1987 and 1996, Korean FDI witnessed spectacular rates of growth, which saw annual levels of investment increase tenfold, from US$320 million in 1987 to US$3.5 billion in 1996 (Bank of Korea 1997: 3).

The increase in Korean FDI in Europe[2] was particularly rapid: cumulative investment, which was below the US$70 million mark in 1986, soared to US$3.1 billion by the end of 1996, with an average annual growth rate of 61.1 per cent (in value terms) in the 1990s (Bank of Korea 1997: 20). Investment throughout Europe was dominated by companies in the fabricated metals sector (including electronic and electrical products, automobiles and ships) which accounted for 71.4 per cent of cumulative manufacturing FDI in the region as of the end of 1996. Within the fabricated metals sector, investment was dominated by the Korean consumer electronics giants, Samsung Electronics, LG, and Daewoo Electronics (Bank of Korea 1997: 95). Korean investors, who had focused their attention on western Europe in the 1970s and 1980s, began to advance into the countries of central and eastern Europe in the 1990s, as South Korea expanded its diplomatic relations at the end of the Cold War era. One particularly notable fact concerning Korean FDI in Europe is that almost 90 per cent of Korean FDI in the region took place between 1992 and 1996, with more than 73 per cent invested between 1994 and 1996 (Bank of Korea 1997: 26–27).

Since the late 1980s, many Korean scholars have published studies that use empirical data to analyze the motivations of individual Korean companies for their investment in Europe (Lee 1989, Yun 1993, Byun

1994, Yu 1994, Shin 1995, Park 1996, Nam 1997 *et al*). Although there has been considerable interest in the micro-level aspects of Korean FDI in Europe, rather less attention has been paid to the theoretical issues underlying much of this investment, and the broader macroeconomic context in which it has taken place.

Established FDI theories put forward by western and Japanese scholars were developed to explain the flows of investment from more advanced to less advanced or developing countries. It can be argued, therefore, that they have limited applicability in explaining investment that flows in the other direction: that is, reverse direct investment (RDI: *yŏkt'uja* in Korean)[3]. Korean consumer electronics investment in the industrialized countries of western Europe is a clear case of RDI and a phenomenon that has generated much debate among Korean scholars in recent years.

A survey of the Korean literature relating to FDI reveals a wide-ranging debate on the issues of Korean overseas investment in general and reverse direct investment in particular. Furthermore, there has been widespread criticism among Korean scholars of the FDI theories put forward by western and Japanese scholars. While some Korean writers have gone no further than presenting a critique of these theories, a few have developed their own RDI models, often incorporating elements from western theories. However, as the authors themselves admit, none of the theories developed offers a full explanation of the RDI phenomenon – the debate continues.

The purpose of this book is to move that debate forward by analyzing Korean consumer electronics investment in central, eastern and western Europe from a theoretical/macroeconomic perspective. This perspective synthesizes elements of western and Korean FDI theory: specifically, Knickerbocker's Oligopolistic Reaction Theory, internalization theory, and the concepts of 'involuntary internationalization' (Yun 1993: 242) and 'structural irreversibility' (Jun 1989: 182). According to Knickerbocker, companies operating in an oligopolistic industry (such as the Korean consumer electronics industry) will exhibit 'follow-the-leader' patterns of FDI in order to maintain the *status quo*. Yun's extension of internalization theory (which states that, faced with imperfect external markets, firms will create internal markets as a means of exploiting their assets) highlighted the internalization of adverse *external* factors such as trade friction and threats to export markets, rather than firm-specific assets. Finally, in developing his Structural Market Defence Theory, Jun incorporated the impact upon FDI activity of irreversible policy decisions taken by the mother company in Korea.

In this book, the applicability of oligopolistic reaction theory to Korean investment throughout Europe is tested using data published by the Electronics Industries Association of Korea, and a framework within which Korean FDI and RDI can be understood is created. Drawing on and extending the concepts put forward by Yun and Jun, this framework incorporates changes in the global and domestic business environment, developments in the political economy of East Asia and Europe, the structural characteristics of the Korean consumer electronics industry, and Korean government policy, both economic and diplomatic. The key question to be answered is: did these factors combine to push Korean consumer electronics companies overseas at a earlier stage than was theoretically feasible or even economically viable? In other words, has Korean consumer electronics investment in Europe truly been a case of 'involuntary internationalization'?

This survey of FDI by Korean consumer electronics companies comprises eight chapters. Chapter 1 offers an overview of the major western and Japanese theories of foreign direct investment. In Chapter 2, the Korean perspective on the FDI debate is analyzed, highlighting the difficulties that are inherent in applying western and Japanese theories to Korean RDI. In this chapter, the works of a number of eminent Korean scholars in the field are presented in English for the first time.[4] The economic context in which FDI has taken place and Korea's industrial structure are analyzed in Chapter 3; the focus is narrowed in Chapter 4, which discusses the development and characteristics of the Korean consumer electronics industry.

The analysis of patterns of Korean FDI in Europe begins with a chronological overview of Korean global FDI (Chapter 5), identifying the changes in the forms and flow of investment that have occurred since the late 1960s. The history of the development of relations between Korea and Europe (Chapter 6) includes an analysis of the Korean view of European integration and a brief overview of the development of relations between Korea and eastern Europe. Chapter 7 analyzes patterns of Korean FDI in Europe within the context of this evolving relationship, focusing on FDI by Korean consumer electronics companies. In Chapter 8, Korean FDI data is analyzed to determine whether there is evidence of oligopolistic reaction in Korean consumer electronics investment in Europe. A theoretical/macro-economic framework is then constructed within which all forms of Korean investment in Europe can be understood. Finally, the conclusions are summarized and areas for future research are suggested.

The cut-off point for the doctoral thesis on which this book is based (31 December 1996) was defined by the economic crisis that came to a head in November 1997, when the ROK government sought assistance from the International Monetary Fund. The 'IMF crisis' and Korea's response to it in terms of economic reform and corporate restructuring will have a profound impact both on government policy regarding outward investment and on existing FDI projects in Europe in terms of consolidation or withdrawal. The final section of the book provides an update on Korea's economic performance and FDI activity since the 1997 crisis.

Chapter One

A Historical Overview of the Development of the Theory of Foreign Direct Investment and the Multinational Enterprise

Over the past five decades, western and Japanese scholars have sought to find a theoretical explanation for the phenomenon of foreign direct investment (FDI) and the rise of the multinational enterprise (MNE). Given the low level of FDI activity by firms from developing countries in the 1950s and 1960s, it was inevitable that the studies published and the theories developed should focus on investment by companies from industrialized countries. However, as Heenan and Keegan observed in 1977:

> Third world multinationalism, only yesterday an apparent contradiction in terms, is now a serious force in the development process. Until the postwar emergence of Japan, we restricted our thinking to multinationals based in North America and Western Europe. Inevitably, though, one nation's competitive dominance passes to another. Today the Japanese, tomorrow the Brazilians, the Mexicans, the Koreans, and others (Heenan and Keegan 1977: 106).

Since the late 1960s, Korean companies have been establishing sales and production facilities throughout the world, investing in industrialized, less developed and developing countries. This chapter traces the development of the major western and Japanese theories of foreign direct investment and the multinational enterprise (including theories of developing country FDI) and highlights their limitations in explaining the full spectrum of Korean FDI activity.

EARLY THEORIES OF FOREIGN PRODUCTION

Prior to the 1950s, attempts to explain the phenomenon of foreign production were based on neoclassical theories of international capital movements and trade, and took the form of country studies that analyzed the advantages offered to investors by the host nation (Tolentino 1993: 31).

However, these theories were unable to provide a satisfactory explanation of the nature and patterns of foreign direct investment. The theory of international capital movements proved to be inappropriate as FDI differs from indirect or portfolio investment in a number of important respects.

Firstly, in the case of indirect (portfolio) investment, when the investor purchases the securities, ownership is transferred from the seller to the buyer. However, when a direct investment is made, the investor enjoys some degree of managerial control over the resources being transferred. Secondly, direct investment results in the transfer of both financial and non-financial resources, which may include technology and management skills. Finally, if the investor raises funds in the capital markets of the host country, direct investment can take place without the transfer of any funds (Dunning 1981b: 76).

The principal shortcoming of international trade theory in explaining FDI is that it takes as one of its basic assumptions the existence of perfect factor and goods markets. In other words, the theory assumes that all firms have equal access to means of production and production inputs (including materials and personnel), and manufacture standard goods that are traded among a number of buyers and sellers who all possess perfect knowledge of prices. However, if goods and factor markets were perfect, there would be no incentive for firms to undertake the risk and expense of establishing an operation overseas.

This is because all enterprises operating outside their home environment encounter a 'cost of foreignness': that is, factors that work against the interests of foreign firms and to the benefit of local companies. These factors may include a lack of familiarity with the local environment, business culture, language and customs, additional overheads and communication costs, the employment of expensive expatriate managers, host government policies that work to the disadvantage of foreign firms, consumer preference for local brands and so forth.

One of the basic premises of the modern theory of FDI and the MNE suggests, then, that foreign production occurs in response to imperfections in goods and factor markets. However, in order to overcome the cost of foreignness, the investing firm must have acquired and developed a compensating advantage, enabling it to operate successfully and compete with actual and potential rivals in the host market.

THE 1960s AND 1970s

In the 1950s and 1960s, western economists attempted to explain the growth of American investment in western Europe after the Second World War. Scholars addressed two basic questions: why were American firms choosing to manufacture in Europe rather than in the United States, and how were they able to compete with local firms, given the cost of foreignness (Casson 1986:

43)? The two main approaches that were developed in response to these questions were based on location theory, which considers the reasons underlying a firm's choice of host country for its overseas investment, and industrial organization theory, which seeks to explain how it is that a multinational firm can compete successfully with local firms in a given market.

The Industrial Organization-Competitive Advantage Approach

In 1960, Stephen Hymer submitted his doctoral thesis, entitled *The international operations of national firms: A study of direct foreign investment*, which developed earlier work on industrial organization theory carried out by Barlow, Southard and Dunning. Hymer argued that firms operating in imperfect markets can develop advantages that give them a competitive edge over rival firms, both in domestic and foreign markets (Hymer 1976: 34–36). As mentioned above, the underlying assumption is that any firm wishing to overcome the cost of foreignness and compete with local firms in a foreign market has to possess a compensating advantage that is (preferably) specific to the company and easily transferable across national boundaries. Buckley noted that firms possessing this type of advantage enjoy a temporary monopoly while their competitors endeavour to develop the same advantage (Buckley 1983: 39).

Kindleberger concurred with the view that FDI cannot exist in a world of perfect competition and perfect markets, and identified four types of market imperfections. These include imperfections in goods and factor markets due to product differentiation, special skills and knowledge, and unequal access to resources and production factors. The other imperfections are internal and external economies of scale that can be exploited through horizontal and vertical integration, and government-imposed limitations on market entry or production (Kindleberger 1969: 13).

The firm-specific advantages identified by Hymer and Kindleberger include advanced technology, superior managerial, administrative and marketing skills, and access to low-cost funding due to the financial strength or credit rating of the parent firm, interest and exchange rate differentials and so forth. Firms that rely on a stable supply of competitively priced raw materials and intermediate inputs and acquire exclusive rights to the development, processing and distribution of those materials will enjoy an advantage over their rivals. Finally, a large-scale firm will benefit from economies of scale and scope, and from the establishment of extensive distribution and marketing networks.

The Product Life Cycle Model

Another important development in the theory of FDI in the 1960s was Raymond Vernon's application of the concept of the product life cycle to

Internalization Theory

The concept of internalization was summarized by Peter Buckley in 1992:[2]

> Internalization is about imperfections in intermediate product markets. Intermediate products flow between activities within the production sector. Market imperfections generate transaction costs and these costs are often minimized for the sector as a whole by bringing interdependent activities under common ownership and control (Buckley 1992: 4).

The theory of internalization, developed in the 1970s, focused on the contrast between external markets and a firm's own internal market. Given that goods and factor markets are imperfect, it is difficult for MNEs to use external markets to recoup their investment in research and development (R&D) and the acquisition of technology, information, and skills. Difficulties in establishing a price for such intangibles as knowledge may prompt a firm to create an internal market through which it can exploit its assets. By internalizing the market, the firm can also deny potential competitors access to the knowledge and thus enhance its competitive position in the short term: that is, until rival firms acquire or develop the knowledge themselves.

In 1975, Williamson had put forward the view that external and internal markets (the market and the firm) are alternative means of selling or exploiting a firm's assets, and noted that the choice between the two is made on the basis of relative efficiency. According to Williamson, the relative inefficiency of external markets *vis-à-vis* the internal market derives from the presence of transaction costs, such as the costs of negotiation (Tolentino 1993: 41). The higher the costs of using the external market, the greater will be the incentive to internalize the market, either locally or overseas.

Working independently of Hymer and Williamson, Peter Buckley and Mark Casson of the University of Reading analyzed existing FDI theories and developed their own theory of the MNE in 1976. They criticized orthodox theories of production and trade on the grounds that these approaches have been superseded by changes in business organizations. Buckley and Casson argued that a firm's activities now extend beyond simple production to include a wide range of interdependent activities including R&D, marketing, training, raising of finance, and the management of financial assets. These activities are linked by flows of intermediate products, which include knowledge, technology, and production and sales experience, as well as materials, parts and components, and semi-finished goods. The efficient coordination of these activities is a prerequisite for business success and, given the difficulty in organizing external markets for some of these intermediate products, there is an incentive to bypass them by creating an internal market, provided that the benefits of so doing outweigh the costs.

Although the creation of internalization theory was clearly an important development in the search for a more holistic approach to the theory of FDI, the new theory had its own limitations. Just as the possession of a firm-specific advantage was a necessary but not a sufficient reason for a firm to engage in FDI, the ability to internalize transactions was not, in itself, sufficient justification for investment overseas.

Dunning's Eclectic Paradigm of International Production

The Eclectic Paradigm of International Production developed by John Dunning represented a major step forward in the search for an all-encompassing theory of FDI. In a monograph published in 1988, Dunning explained his choice of the word 'eclectic' as aiming to convey the idea that FDI or MNE activity is best explained by a 'synthesis of elements from a number of theories, creating a general framework within which FDI can be understood' (Dunning 1988a: 41). According to Dunning, three types of advantage must exist for a firm to engage in FDI: ownership-specific advantages, location-specific advantages, and internalization incentive advantages.

A firm's ownership-specific advantages may derive from its size and market position (with the advantage increasing if the firm enjoys a monopolistic or oligopolistic position within the market), its ability to develop new products or processes, the possession of proprietary technology, trade marks, brand names, and so forth. A firm's managerial, technical and research personnel, its production management, administration and marketing systems, and its access to production factors are also sources of advantage. The firm may also enjoy advantages stemming from an affiliation with a parent company that can provide access to finance, administrative, managerial, R&D and marketing capabilities, and economies of joint production, marketing, distribution and so forth.

Location-specific advantages offered by a host country include access to local and regional markets, availability of low-cost production factors, competitive transportation and communication costs, the opportunity to circumvent import restrictions, and investment incentives offered by the government of the host country. The internalization incentive advantages identified by Dunning offer protection against or opportunities to exploit market failure, and include the reduction in costs associated with transactions in the external market. An additional incentive for FDI will be the opportunity to control sources of production inputs or sales outlets that might otherwise by exploited by rival firms (Dunning 1981b: 80–82).

Kojima's Macroeconomic Approach

In the late 1970s, Kojima Kiyoshi analyzed the resource endowments of the United States and Japan and their impact on overseas investment by firms from those countries. Kojima concluded that Japanese firms engaged in 'trade-oriented investment' (later re-named 'development-oriented investment') whereas investment by American firms could be characterized as 'anti-trade oriented'. In the 1970s, the main focus of Japanese FDI was investment in resource development or small-scale, labour-intensive production in developing countries. Kojima argued that this investment could be regarded as 'trade-oriented' or 'development-oriented' in that it led to an expansion of bilateral trade and benefitted both Japan and the host country. By engaging in FDI, Japan was able to secure supplies of scarce natural resources and low cost labour, enhance its industrial structure, and expand its exports. The host (developing) country benefited from assistance in establishing a strong industrial base and creating its own competitive advantages, as well as gaining access to technology appropriate to its stage of economic development.

In contrast, American investment was focused on sectors in which US firms enjoyed an oligopolistic or innovative advantage. (Typically, these were the more advanced, capital- and technology-intensive sectors, such as computers, chemicals, and pharmaceuticals.) These firms engaged in investment in order to exploit their competitive advantage (or 'monopolistic absolute advantage', as Kojima termed it). This type of investment, Kojima argued, was a substitute for exports and served merely to advance the interest of the multinational corporation (Kojima 1986: 380–386).

MOTIVATIONS FOR INVESTMENT BY MNEs FROM INDUSTRIALIZED COUNTRIES

According to Dunning (1993b), there are four main categories of motivation for foreign production or foreign direct investment: resource-seeking, market-seeking, efficiency-seeking and strategic asset- or capability-seeking. A firm engaging in FDI may be influenced by one or more of these, and the motivations for foreign production by that firm may change over time. Resource-seeking investors will move overseas to secure a stable supply of necessary resources, which are either expensive, in short supply or unavailable at home. These are commonly raw materials or energy sources, but can also include labour, technology, management skills and marketing know-how. The acquisition of these resources will increase the firm's profitability by lowering production costs, and also enhance its competitiveness in domestic and foreign markets.

Market-seeking investors move overseas to defend market positions already established through exporting, or to open up new markets for their goods and services in the host country or neighbouring countries. The purpose of the investment may be to take advantage of incentives offered by the host government or to circumvent current or proposed trade restrictions, including both tariff and non-tariff barriers. Equally, the investor may be seeking to reduce production and transaction costs by manufacturing locally. For some companies, such as manufacturers of parts and components, a move overseas by a major client may prompt a similar action in the interests of maintaining and expanding the business relationship. The firm may also take the decision to establish a position in the market in order to develop a better understanding of local requirements, consumer tastes and business culture.

Corporate efforts to rationalize these resource-seeking or market-seeking investments in order to benefit from economies of scale and scope will result in efficiency-seeking investments. These investments aim to maximize the benefits derived from producing in a small number of countries – each with its own advantages in terms of factor and resource endowments, government incentives and so forth – in order to service a large number of markets. Finally, firms engaging in strategic asset-seeking investment do so with the long-term aim of maintaining or enhancing the firm's international competitiveness (Dunning 1993b: 54–62).

THEORIES OF DEVELOPING COUNTRY FDI

The substantial amount of literature and data on foreign direct investment by MNEs from advanced countries reflects the heavy concentration of outward investment activity among those nations. It is only in the past two decades that firms from developing countries have emerged as outward investors, as their trading activities have increased, their manufacturing capabilities have grown, and their service sectors have begun to develop. (As was noted in the introduction, the Koreans themselves only became active investors in the latter half of the 1980s.) Changes in these firms' industrial structure, their factor endowments and the global economic environment have created a need for MNEs from some developing countries to move overseas. These developments fuelled a debate on the necessity for an independent theory of FDI from developing countries. Did this investment have its own unique nature and patterns, or would these characteristics, given time, converge with those of FDI by industrialized countries?

Two major theories of developing country FDI are presented below: Dunning's concept of an investment development cycle and Wells' combination of the analysis of firm-specific advantages with Vernon's product life cycle theory.

Dunning's Investment Development Cycle

The Investment Development Cycle (IDC), which was first presented at a Conference of Third World Multinationals in Honolulu in 1979, was used by Dunning to explain the changing levels of inward and outward investment experienced by countries during the process of economic development. According to Dunning, the level of FDI activity seen in a given country (both in terms of inward and outward investment) will vary according to a number of factors, including its level of economic development, its factor endowments (natural resources, labour and so forth), and its political and economic systems (Dunning 1988b: 15–17). Changes in the country's level of economic development will affect both the advantages possessed by local companies and the attractions of the country for potential inward investors. According to Dunning, the net outward investment (NOI) position of a country (the difference between levels of inward and outward investment) is a function of that country's stage of development as measured by per capita gross national product (Dunning 1993b: 88).

Dunning identified five stages of investment activity during a country's economic development. In the first stage (the pre-industrialization phase), because the advantages possessed by local firms are too weak to make outward FDI feasible, companies focus on exports. Similarly, at this stage, the country's location-specific advantages are insufficient to attract inward investment, and its political, commercial, legal and technological infrastructure is not sufficiently developed to be able to offer the level of support required by foreign investors. During this phase, the country's NOI level is negative or close to zero.

However, as the economy develops and the infrastructure improves, foreign investors enter the market to gain access to natural resources, to establish labour-intensive production facilities, to produce goods for the local market, or to exploit some other investment opportunity. Thus, inward investment expands, but outward investment remains at a low level, as domestic firms have not yet developed sufficient firm-specific advantages to advance into foreign markets. The NOI level becomes increasingly negative during this stage, as inward investment exceeds outward investment.

In the third stage, as the country becomes more developed, foreign investors may be attracted by the potential of the local market or the availability of skilled labour, rather than by access to low-cost labour or raw materials. The decline in availability of these production inputs will also encourage local firms to engage in FDI in order to seek markets and resources (including advanced technology) overseas. Although NOI is still negative at this stage, the gap between inward and outward investment is reduced, due either to a fall in inward FDI or an increase in overseas investment activity by local firms.

During the fourth stage, NOI is positive and rising as the economy reaches a mature phase and the strength of the OLI advantages possessed by local firms prompts higher levels of outward investment, which now exceeds inward investment. The fifth stage of the investment development cycle was added by Dunning at a later date to reflect the increase in cross investments among major industrialized nations. This results in a reduction in NOI levels as inward investment once again begins to outpace outward FDI. In contrast to the other stages of the cycle, in which the country's own stage of economic development is a crucial factor, the final stage depends more on the country's economic position in relation to that of other countries (Tolentino 1993: 87–92).

Louis T. Wells – Appropriate Technology and Appropriate Products

Writing in 1977, Wells viewed the internationalization of firms from developing countries as a stage in the product life cycle or the technological gap model described by Hufbauer in 1965. Hufbauer put forward the view that countries fall into a pecking order based on their ability to manufacture a given product. Countries possessing the necessary manufacturing facilities will export to countries further down the pecking order that do not. The development of FDI follows the same pattern, with activity led in the first instance by industrialized countries and developing countries engaging in FDI at a later point in time (Tolentino 1993: 74).

Wells identified three types of advantage for firms from developing countries: the acquisition of technology suited to small scale manufacturing, the ability to produce goods particularly suited to a regional or ethnic group, and an established sales presence in the market. Among these, he identified technology as the key advantage in the FDI activities of firms from developing countries (Wells 1983: 1–4, 19–65). Wells found that these firms tend to operate mass production facilities relying on mature, standardized technology imported from industrialized countries. The prospect of creating excess capacity and facing high fixed costs deters MNEs from industrialized countries from using this mature technology in developing countries. However, firms in those markets have already adapted the technology and scaled it down for use locally, thus creating an advantage that can be exploited in other developing countries. In those countries, labour-intensive, flexible technology is needed to allow short production runs of a wide variety of products.

In addition to adapting imported technology, some investors from developing countries begin to manufacture related parts and components locally, either because this is a low-cost alternative to importing, or as a response to import restrictions imposed by their own government in order to curb the country's current account deficit. Again, the parts and components developed locally can be exported to or exploited in less developed or developing countries (Wells 1977: 138–145).

Looking at patterns of foreign direct investment, Wells observed that multinational companies from developing countries have a tendency to establish foreign manufacturing affiliates in neighbouring countries at a level of economic development similar to or (more often) lower than that of their own. Wells concluded that advantages enjoyed by developing country multinationals derive from their ability to adapt imported technology to local conditions rather than from their products *per se*. These firms are able to develop small-scale, flexible, multi-function technology suited to developing country markets and conditions (Wells 1983: 1–4, 19–65). In 1993, Wells wrote that Vernon's product life cycle model can be used to explain the phenomenon of FDI from developing countries. Just as Vernon's theory argues that production moves down the ladder of development as products mature, so FDI flows from the richer to the poorer developing countries (Wells 1993: 183).

MOTIVATIONS FOR INVESTMENT BY MNEs FROM DEVELOPING COUNTRIES

Writing in 1977, Wells identified a number of factors that encourage firms from developing countries to engage in FDI, including government policies and measures aimed at promoting outward investment, the limitations of the domestic market (due to insufficient current or potential demand for their products), and expectations that transferring technology overseas will bring greater financial benefits than diversifying into new areas at home. In addition, firms may invest to secure supplies of raw materials, establish bases from which to promote exports, circumvent import restrictions, or develop products suited to a particular market or ethnic consumer group (Wells 1977: 136).

An empirical study published by Wells in 1983 revealed that most developing country firms invest overseas in order to defend their export markets in the face of actual or potential trade restrictions. By investing in countries not subject to import restrictions and quotas, the firms are able to protect and even expand their export markets. Investing in countries with low-cost labour enables them to reduce their production costs and enhance their price competitiveness. Some firms establish foreign production bases as a means of accessing a third market through a 'back door' route: a good example of this is investment in Mexico to gain access to the North American market.

Wells discovered some evidence that the existence of ethnic groups in the host market acts as a motivation for investment, which may be made on the basis of information received from expatriates or family members living there. Other firms, faced with the prospect of political and economic instability in their home market, turn to foreign direct investment as a means of hedging that risk. Wells concluded that firms will only take the risks associated with FDI after all other avenues for defending the export market have been explored, or when intelligence received from the market indicates that the risks of investment will be limited (Wells 1983: 67–83).

CONCLUSIONS

This chapter has provided an overview of the major western and Japanese foreign direct investment theories, in terms of investment by MNEs from both industrialized and developing countries. Although the analysis of and debate on FDI theory will surely continue in the secondary literature for many years to come, there now appears to be a general consensus among western scholars that a core theory has been established on the basis of which future research can be carried out into the activities of all MNEs, be they from industrialized or developing countries (Buckley 1990: 657–665).

However, as will be seen in Chapter 2, Korean scholars have expressed doubts about the ability of these theories to explain all the forms of Korean overseas investment. A preliminary survey of the theories does, indeed, suggest that there may be limitations as regards their applicability to global Korean FDI activity. This raises a number of questions that will be addressed in the following chapters. Firstly, a key premise of most of the theories discussed above is the possession of a transferable and (preferably) firm-specific advantage that can be used to offset the cost of foreignness facing companies engaging in FDI. This advantage can take the form of advanced technology, trade marks, brand names and product differentiation, marketing and managerial skills, extensive distribution networks, access to financing on favourable terms, and the availability of low-cost production factors.

When companies are investing in countries at an earlier stage of economic development (that is, further down Hufbauer's 'pecking order'), additional advantages include the possession of mature, standardized technology that has been adapted for use in developing countries, access to industrialized country markets, and ethnic ties with the host market. This raises a fundamental question: have Korean consumer electronics companies invested in Europe on the basis of a firm-specific monopolistic advantage?

Secondly, in addition to ownership advantages that can be exploited through the creation of internal markets, Dunning's Eclectic Paradigm requires the possession of location-specific advantages, which include market access and the circumvention of trade barriers. To what extent have location-specific advantages motivated Korean investment in Europe? Thirdly, the product life cycle theory describes a sequential process by which innovators move from domestic production to exports and, finally, to overseas production. Have Korean consumer electronics companies invested in Europe on the basis of their innovation within the industry? Finally, Knickerbocker suggests that firms within an oligopolistic industry will exhibit follow-the-leader patterns of FDI in order to maintain the *status quo*. Can these patterns of investment be identified in Korean FDI in Europe?

In the next chapter, the Korean perspective on the foreign direct investment debate will be presented and analyzed, both in terms of Korean scholars' critique of the theories discussed in this chapter, and in the light of the original contributions made by Korean scholars in the field. This analysis will offer an additional perspective on the ability of all these theories to explain the phenomenon of Korean consumer electronics FDI in Europe.

Chapter Two

Foreign Direct Investment Theory: The Korean Perspective

An examination of the principal western and Japanese theories of foreign direct investment has raised a number of questions concerning their applicability to patterns of Korean investment. The purpose of this chapter is two-fold: to present and analyze Korean contributions to the FDI debate, and to consider to what extent these contributions can be said to have moved the debate forward.

A survey of the Korean literature reveals that, while there has been widespread criticism of the inability of existing western and Japanese theories to explain fully the forms and flows of Korean global investment, only a few Korean writers have made their own, original contributions to the debate. The main criticism of the established theories of FDI put forward by Hymer, Dunning, Wells *et al* is that they can only give a satisfactory explanation of certain phases of Korean investment overseas. It is argued that, having been developed to explain the flows of investment from more advanced to less advanced countries, they have limited applicability in explaining investment flowing in the other direction: that is, reverse direct investment (RDI).

Between 1968 and 1980, Korean foreign direct investment was focused mainly on resource development and labour-intensive manufacturing in neighbouring developing countries and less developed countries. During this period, Korean investment in North America and Europe accounted for less than 30 per cent of Korea's global FDI (Bank of Korea 1996: 26–27). Investing in developing countries, Korean investors successfully exploited their firm-specific advantages, including the possession of simple, standardized and small-scale technology that had been adapted to suit local market conditions, and experience of operating in countries at a similar or earlier stage of economic development (See Chapter 5). As established theories were able to offer an adequate explanation for the majority of Korean FDI activity in the 1960s and 1970s, there was little need for an independent theory of Korean FDI at this stage.

However, during the 1980s, Korean corporations began to establish production bases in industrialized countries such as the United States, the United Kingdom, and West Germany. They invested in the absence of any clear firm-specific monopolistic advantage which would enable them to overcome the cost of foreignness and compete with host country firms and other MNEs on an equal basis. This phenomenon of reverse direct investment could not be explained by existing theories, which assumed that such an advantage was an essential precondition for successful investment overseas.

This section presents a representative sample of Korean approaches to FDI theory, with specific reference to theories relating to reverse direct investment. The literature, presented in chronological order, includes monographs and articles that offer general theoretical studies of FDI and the MNE, and postgraduate dissertations and theses that survey existing FDI theory and, in some cases, put forward new theoretical models. As far as the author is aware, this is the first time that the majority of these works have been presented in the English language. The exceptions are Koo (1985), Jun (1987), Choi (1991), Jeon (1992), and theses submitted by Korean postgraduates studying in the United Kingdom.

Although a considerable number of works have been published by Korean scholars on the subject of FDI theory, most have gone no further than providing a standard summary and critique of existing FDI theories. The work carried out by Yun Tong-jin (1986, 1993) in his master's degree dissertation and doctoral thesis constitutes an original contribution to the debate and, as such, is discussed at the end of the chapter along with other theories developed by Jun Yongwook (1989) and An Se-yŏng (1995).

A CHRONOLOGICAL SURVEY OF THE KOREAN LITERATURE

Writing in 1985, Koo Bohn-young noted the partial applicability of existing western and Japanese theories to Korean FDI, and tested Dunning's Investment Development Cycle against Korean economic and investment data collected between 1962 and 1982. Although the data analyzed did reveal a relationship between Korea's net outward investment position and its level of economic development, Koo noted that the levels of investment were affected as much by the policies implemented by the ROK government as by changes in Korea's country- or firm-specific advantages. In particular, Koo highlighted the importance of Korea's balance of payments situation in determining government policy on FDI. In periods when the trade and current accounts were in surplus, outward investment would be encouraged as a means of averting protectionist measures by trading partners. Equally, low levels of inward FDI might be due to government restrictions on foreign investment in Korea (Koo 1985: 290–302).

Pan Pyŏng-gil (1985, 1996) recognized the phenomenon of reverse investment with his classification of three patterns of developing country

FDI (DCFDI): investment flows to a country at an earlier stage of development, investment in a country at a similar stage of development, and investment in industrialized countries (Pan 1985: 143). However, in his consideration of the characteristics of and motivations for DCFDI, Pan focused on the first two patterns and his analysis did not include reverse direct investment, which was in its infancy at the time.

Pan noted the existence of two contrasting views on investment by firms from developing countries. While some analysts saw many similarities between DCFDI and investment by MNEs from the United States and Japan in the early stages of their advance overseas, others took the view that DCFDI must be seen as a completely new form of direct investment. The latter view was based on a number of considerations specific to DCFDI: the fact that it tends to be small in scale, that investment is usually greatly influenced by the policies of the home government (such as controls over foreign exchange and capital movements), and that there is a tendency to focus on specific industrial sectors, host countries and markets favoured by the home government (Pan 1996: 171).

Pan identified a number of special features of DCFDI. Firstly, there is a strong tendency for developing country firms to invest in other (neighbouring) developing countries. Secondly, as Wells suggested, the technology and products transferred by the means of FDI among firms from developing countries tend to be 'relatively more appropriate' to local conditions than those transferred through investment by firms from industrialized countries. Thirdly, investment tends to take the form of joint ventures; this is not only because most developing country host nations require that inward investment take this form, but also because investors from developing countries have a preference for joint ventures over wholly-owned investment, often reflecting a desire to spread the risk of investment. Once again, Pan's focus was on investment in developing and lesser developing countries, rather than reverse investment (Pan 1985: 144–145).

In 1986, Wŏn Chong-gŭn carried out a study of reverse direct investment in the United States by Korean consumer electronics firms that became a landmark work in the Korean study of FDI. Rather than developing a new theory of FDI to explain this classic example of reverse investment, Wŏn used a combination of Dunning's Eclectic Theory, the theory of the growth of the firm, and oligopolistic reaction theory to explain a specific case of reverse direct investment within the consumer electronics industry (See Figure 2.1).

Wŏn demonstrated that Korean consumer electronics firms had invested in the United States as a response to trade barriers and investment incentives in the market. Although they lacked a clear firm-specific advantage, they were able to internalize the comparative advantage of importing low-priced parts for assembly from Korea. By investing in the United States, Korean

Figure 2.1 Wŏn Chong-gŭn's Theoretical Explanation of Reverse Direct Investment

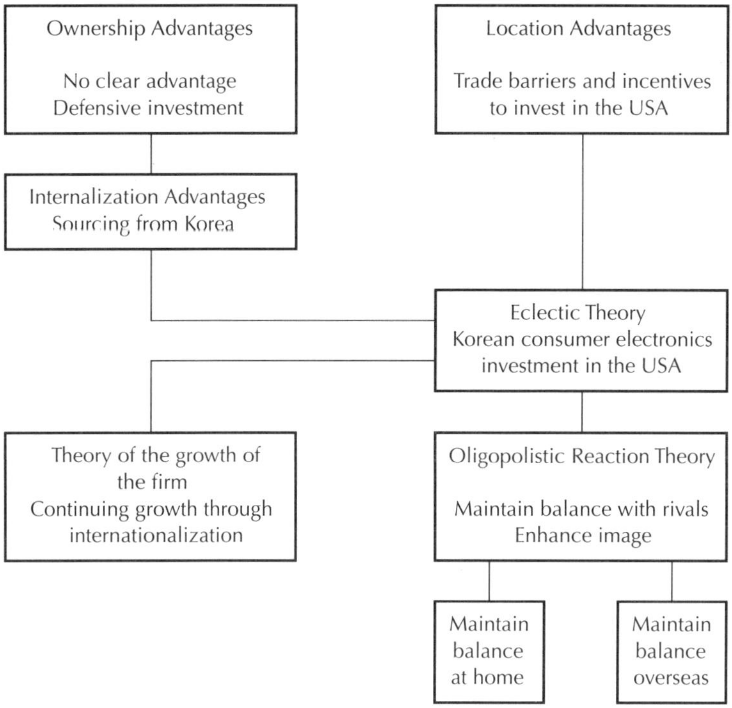

Source: Wŏn Chong-gŭn 1986: 175.

consumer electronics firms were seeking to promote their continuing growth and maintain the oligopolistic balance between themselves and their rivals, at home and overseas (Wŏn 1986: 175).

Wŏn concluded that, while it was possible to explain many aspects of RDI using a combination of these theories, the model was unable to explain in full cases of investment in industrialized country markets by firms lacking a clear firm-specific monopolistic advantage (Wŏn 1986: 207–208). As Wŏn made no reference to existing Korean theories of reverse direct investment or original Korean contributions to the FDI debate, either in this work or in the major bibliographical survey of FDI theory that he published in 1987, it may be inferred that no such theory was in existence by the mid-1980s.

Wŏn's 1987 bibliographical survey was the most comprehensive treatment of western and Japanese FDI theory by a Korean writer of its time, and it remains one of the classic works on the subject in the Korean language. In his article, Wŏn evaluated the theories existing in the late 1980s, covering the major western theories and the contributions made by Kojima and Ozawa. Wŏn noted that Dunning's Investment Development Cycle was accepted as being the most general theory to date, and also made reference to Wells' work in his discussion of the Pecking Order Theory, originally developed by Hufbauer (Wŏn 1987: 85–91). Aside from these references, no mention was made of DCFDI theory or views on the emerging phenomenon of reverse direct investment.

In his conclusion, Wŏn wrote that none of the existing theories could explain the complex phenomenon of FDI in a consistent manner, and put forward the view that the various theories should be considered as complementing rather than competing with each other. Although he wrote of the 'urgent need' to develop a unified theory model, Wŏn did not undertake that task himself (Wŏn 1987: 74, 106).

In one of the few English language articles to have been published on the topic of reverse investment, Jun Yongwook (1987) characterized RDI as 'defensive' and 'immature', with investment being made in order to defend export markets rather than to exploit a firm-specific monopolistic advantage. Given this lack of a clear advantage, it is difficult for conventional FDI theories to explain the phenomenon of Korean investment in countries such as the United States and the industrialized nations of western Europe. According to Jun, such advantages as were possessed by Korean investors in these countries did not occur naturally but were 'squeezed out'. As Wŏn had observed, these firms capitalized on Korea's advantage of low-cost labour by importing key parts and components from home for assembly overseas, thus retaining their advantage of low manufacturing costs and maintaining their price competitiveness (Jun 1987: 91–96, 99–101). However, writing in 1991, Ch'ae Chŏm-gil noted that this advantage was being eroded by the strengthening of local content requirements for foreign production in industrialized countries. This posed a significant threat to the price competitiveness of these Korean investors (Ch'ae 1991: 25).

Shin Ki-t'aek (1990) noted the existence of many problems as far as a theoretical explanation of Korean FDI was concerned, and identified the short history and small scale of Korean outward investment as one of the barriers to a full analysis of the phenomenon. According to the general view of Korean FDI presented by Shin, Korean outward investment in the 1980s could be regarded as an extension of export activity or a method of promoting exports: that is, FDI and exports were complementary. Secondly, FDI had contributed to Korea's acquisition of foreign technology, and, thirdly, Korean firms did not possess the clear monopolistic advantages that were generally perceived to be a necessary condition for FDI, such as superior marketing capabilities, product differentiation, advanced technology or access to finance. Finally, although Korean firms possessed a comparative advantage *vis-à-vis* firms from industrialized countries in terms of low production costs and standardized labour-intensive technology, these were not, generally speaking, significant motives for internalization.

Shin concurred with other Korean scholars in observing that the monopolistic advantage possessed by Korean investors in developing countries derived from a number of factors, including the development of standardized, labour-intensive technology suited to small-scale and low-income markets, the ability to reduce local production costs by employing Korean expatriate labour at overseas plants, and access to materials and intermediate goods imported from Korea. Other important factors were strong support from the home government, local marketing experience gained through exports, and the ability to carry on business in countries with a low level of infrastructure development (Shin 1990: 12–15).

The most recent survey of FDI theory considered here was published in 1996 by Choi Yong-rok. In a work published five years earlier, Choi had observed that Korean firms investing overseas generally lacked ownership advantages such as advanced technology or management skills. According to Choi, ROK government policy had promoted the flow of investment towards countries whose location-specific advantages were derived from low production costs, in an effort to help Korean firms maintain their price competitiveness in international markets. In common with firms from other developing countries, Korean investors had tended to adopt a minimum price strategy, focusing on the lowering of production costs, rather than devoting resources to the enhancement of product quality through research and development (R&D), marketing, distribution and so forth (Choi 1991: 11, 53).

Choi put forward the view that Korean outward investment represented an intermediate case between developed and developing country FDI, and identified four variables common to both types of investment. Choi's basic premise was that the essential variables of developed country FDI are localization advantages, internalization advantages and intangible assets, while the essential variables of developing country FDI are localization advantages, internalization advantages and government policy. Choi's model combined developed and developing country theories by identifying two essential and two additional variables for FDI.

The two essential variables are localization and internalization advantages; without the former, firms will continue to export to the market, and without the latter, they will be unable to compete with local firms. Choi noted that firms from developed countries tend to exploit internalization advantages, whereas their developing country counterparts depend more on push factors at home and pull factors abroad. The two additional variables, intangible assets and government policy, may or may not serve to promote FDI depending on the circumstances of the investor country. These variables provide additional incentives to invest, and strengthen the firm's ability to compete with host country competitors. Choi claimed that, according to his model, any firm can engage in FDI, provided that the necessary conditions of localization and internalization are satisfied (Choi 1991: 49–52).

Writing in 1996, Choi identified six major strands of DCFDI theory, namely internalization theory, localization theory, ethnic tie theory, minor innovation theory, minimum price strategy theory, and political system environment theory. In explaining these six elements, he offered some examples of Korean FDI that could be explained by various theories. It is particularly interesting that reverse investment was mentioned only in reference to localization theory (in terms of the strong attractions of the host environment as an incentive to invest), rather than in connection with the exploitation of monopolistic advantages. Once again, the main focus of the work lay in the consideration of the motives underlying the decision to engage in FDI, rather than the identification of firm-specific advantages that made overseas investment possible.

According to Choi, internalization theory as applied to DCFDI shows that firms engaging in overseas investment do so to in order to free themselves from constraints imposed on them at home, either by the government or by market conditions. The investors seek to achieve economies of scale, enhance their price competitiveness and, ultimately, boost their exports by expanding their presence in overseas markets. (A good example of this kind of FDI is investment by Korean footwear manufacturers in South-east Asian countries with low labour costs.) Firms from industrialized countries, on the other hand, regard overseas investment as a substitute for export activity, and engage in FDI in order to enhance their *non-price* competitiveness.

Localization theory demonstrates that country-specific factors such as the availability of abundant, low-cost labour and access to advanced technology and abundant natural resources are important in attracting investment from both developing and developed countries. These factors have played an important role in promoting Korean investment in labour-intensive manufacturing in South-east Asia, natural resource development projects in resource-rich countries such as Indonesia, and high-tech sectors in Europe (reverse direct investment).

The ethnic tie theory holds that the possibility of exploiting ethnic ties overseas can act as a motivation for FDI, with the added benefit of minimizing the risks involved in the early stages of investment by obtaining information from ethnic groups in the host country. In the early stages of Korean FDI, firms targeted areas of the United States with large populations of emigrant Koreans. A similar phenomenon was seen in the late 1980s and early 1990s as Korean firms invested in north-east China and central Europe, where there are significant pockets of ethnic Korean population from which investors can recruit workers.

The minor innovation theory shows how developing countries import technology from advanced countries and adapt it to suit their own market conditions and those of neighbouring developing countries. According to the minimum price strategy theory, the primary aim of DCFDI is to secure price competitiveness by minimizing all direct and indirect costs associated with production, trade, and investment in marketing, advertising and R&D. Finally, the political system environment theory considers the relative importance of political risk among host nations and factors relating to growth, profitability and stability in the host nation. That is, the decision to invest can be seen as an attempt to avoid economic restrictions imposed by the home government rather than a response to encouragement or incentives to invest overseas (Choi 1996: 53–57).

KOREAN THEORIES OF FOREIGN DIRECT INVESTMENT

This section presents and analyzes the theories developed by Korean writers who have attempted to explain the phenomenon of Korean investment in industrialized countries. The four theories considered here are:

Yun Tong-jin	Integrated Theoretical Model – *T'onghap Model* (1986)
Jun Yongwook	Structural Market Defence Theory – *Kujojŏk Shijang Pangŏron* (1989)
Yun Tong-jin	Post-Product Life Cycle Model – P*'osŭt'ŭ PLC Model* (1993)
An Se-yŏng	Competitive Advantage–Representative Specialization Zone Theory – *Taep'yo T'ŭkhwa Chidae Iron* (1995)

Yun Tong-jin's Integrated Theoretical Model (ITM – *T'onghap Model*)

In a work published in 1986, Yun Tong-jin concurred with other Korean scholars that it was difficult to explain investment by MNEs from developing countries in industrialized countries on the basis of existing theories. Yun developed his own theoretical approach, the Integrated Theoretical Model (ITM) which, he claimed, can offer an explanation of reverse direct investment. Rather than being a completely new model, the ITM is an extension of internalization theory which also incorporates the principal concepts of other existing theories. Yun criticized internalization theory on the grounds that it takes as its basic premise the possession of a firm-specific monopolistic advantage, which is then exploited through the creation of an internal market. Therefore, this theory struggles to explain FDI in industrialized countries which is undertaken by firms from developing countries with no clear firm-specific advantages (Yun 1986: 108).

Working from the basic concepts of internalization theory, Yun wrote that firms create and make use of their own internal markets when transactions in external markets become inefficient or impossible. Factors causing these imperfections and inefficiencies can be classified into two categories: internal and external. Looking first at the internal factors, Yun noted that a company's firm-specific monopolistic advantages commonly derive from the accumulation of knowledge or knowhow concerning production, finance, marketing, personnel management and so forth. However, as these advantages have the nature of a public good, transactions in external markets become inefficient, and so the firm chooses to internalize the advantage, and engages in FDI in order to exploit the knowledge that it possesses.

External factors identified by Yun include imbalances and imperfections in external markets due to government intervention, which may encompass tariff and non-tariff barriers, economic and tax policies, emigration and anti-trust legislation, government support and incentives. In cases where government regulations make external market transactions impossible, the firm will engage in FDI. Yun contrasted the 'positive and voluntary' approach to FDI encouraged by internal factors with the 'passive and negative' approach necessitated by changes in external factors. Whereas the former is a mature form of investment which aims to exploit the firm's own capabilities and assets, the latter is immature investment undertaken as a reaction to changes in the external environment.

The ITM thus makes the case that FDI can occur on the basis of external factors, even in the absence of a clear firm-specific monopolistic advantage, because investors internalize not only firm-specific advantages but also changes in the environment. Yun argued that, in the case of investment in the United States by consumer Korean electronics firms in the 1980s, the decision to invest was based on an analysis showing that the benefits of internalization outweighed the costs. External factors including the erection of trade barriers in the US market were more important than factors internal to the firm, such as the possession or lack of a monopolistic advantage. Using this example, Yun made the case that his model can explain all FDI flows, as the phenomenon of reverse investment can be seen as the process of internalizing external factors (Yun 1986: 109–115).

Jun Yongwook's Structural Market Defence Theory (*Kujojŏk Shijang Pangŏron*)

Writing in 1989, Jun Yongwook argued that Korean investors in industrialized markets either did not possess the firm-specific monopolistic advantage that most theories take to be a prerequisite for FDI or, where such an advantage did exist, it was too weak to offset the cost of foreignness. Jun argued that, by its very nature, reverse direct investment is reactive and defensive, and so is not a choice made on the basis of a careful comparison of the economic feasibility of the various means of entry into a foreign market (that is, exports, licensing and FDI). Rather, it is a response to the imposition of trade restrictions that threaten a major export market.

Another key factor is the irreversibility of decisions already taken by the parent firm. A firm that adopts an export-led strategy and relies on mass production to achieve economies of scale and price competitiveness in foreign markets can ill afford the decline in exports that may follow the imposition of import restrictions in a major market.

A decline in or suspension of export performance would result in lower factory operating rates at home, leading in turn to weaker price competitiveness and a decline in business performance. Unable to switch from its high-volume, low-price strategy or open up new export markets

Figure 2.2 Jun Yongwook's Theory of Structural Market Defence

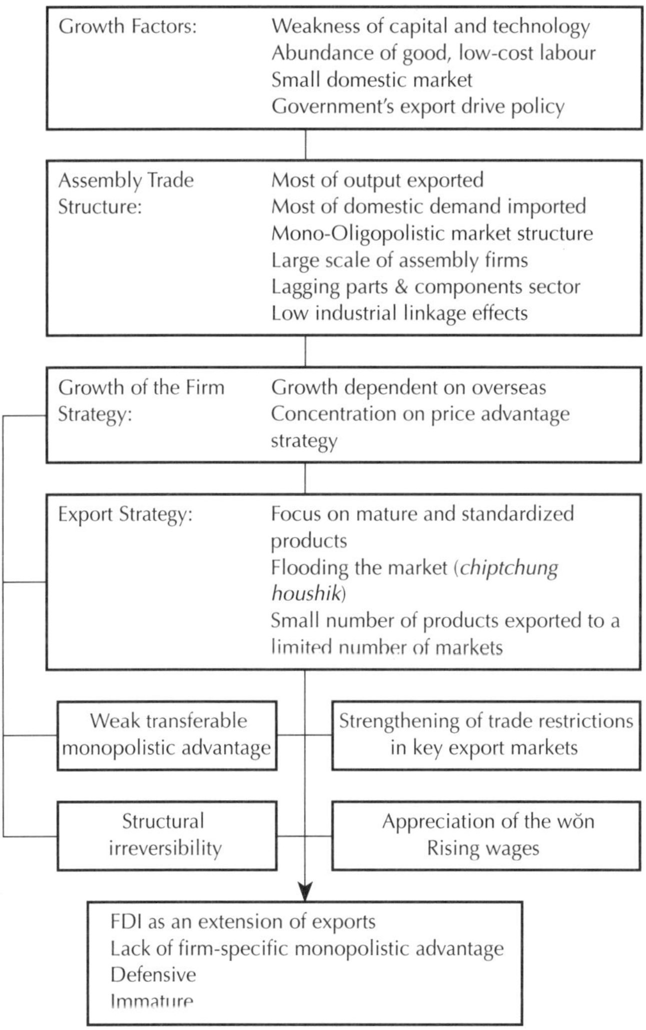

Source: Jun Yongwook 1989: 182.

within a short space of time, the firm has to invest directly in the market, regardless of whether or not it has a firm-specific advantage. Jun concluded that Korean firms had engaged in reverse investment as a matter of survival in the face of changes in the external environment, rather than as a means to exploit a monopolistic advantage (Jun 1989: 182).

Jun characterized Korean FDI as defensive, in that it was undertaken in order to protect export markets, and premature, in that it would not have occurred had there not been a threat to the market. Companies had engaged in reverse direct investment in order to circumvent import restrictions, maintain factory operating ratios and economies of scale at home, enhance their corporate image and brand acceptance among Korean consumers, and to keep pace with rival firms (Jun 1989: 174–178). Jun also considered the impact of Korea's industrial structure on patterns of FDI, identifying a number of factors that encouraged firms to move overseas. These included Korea's weak technology base and lagging parts and components sector, the export strategy adopted by the ROK government (high-volume exports of low-priced goods to a small number of markets), and weak R&D and marketing capabilities.

Jun identified three factors that had enabled Korean firms to invest overseas regardless of the lack of a clear firm-specific advantage: firstly, investors were able to maintain their price competitiveness by restricting their overseas activity to the assembly of parts, components and semi-finished goods imported at low cost from Korea. In this way, they were able to internalize Korea's advantage of low-cost labour. Secondly, investors affiliated with the *chaebŏl* were able to take a long-term view as regards the profitability of a new overseas operation, given that the group as a whole could absorb low profitability or even losses in the early stages of operation. Finally, the firms had adopted and implemented a variety of strategies aimed at minimizing production costs, in order to further enhance their price competitiveness (Jun 1989: 180–182).

Yun Tong-jin's Post-Product Life Cycle Model (P'*osŭt'ŭ PLC Model)*

In 1993, Yun Tong-jin submitted his doctoral thesis, in which he put forward a second theoretical model, which he called the Post-Product Life Cycle Model (P'*osŭt'ŭ PLC Model)*. As its name suggests, the model expands upon Vernon's theory, using seven stages of trade and investment to explain the internationalization process of firms from developing countries. The model also identifies four groups of investors: firms from advanced countries (AC) such as the United States, less-developed advanced countries (LDAC) such as Japan, developing countries (DC) such as Korea, and lesser developing countries (LDC) such as Vietnam. The interesting feature of this theory is that it was developed specifically to explain FDI in advanced countries by Korean consumer electronics firms: that is, RDI within a specific industry.

Figure 2.3 Yun Tong-jin's Post-Product Life Cycle Model: The Seven Stages of Trade and Investment for Korean Consumer Electronics Firms

Stage 1	No demand in the home market
Stage 2	Demand arises and imports begin
Stage 3	Active import substitution: investment by AC firms in DC markets to benefit from low labour costs
Stage 4	Exports to AC markets based on low labour costs and economies of scale
Stage 5	Investment in AC markets, exploiting the advantage of sourcing low-cost parts, components and semi-finished goods from home
Stage 6	Transfer of production to LDC markets due to erosion of comparative advantage
Stage 7	Importing from LDC markets as competitive advantage disappears

Source: Yun Tong-jin 1993: 263.

Yun claimed that this revised theory is able to explain the phenomenon of investment in AC markets by firms from developing countries (i.e. reverse investment), which existing theories had failed to do. The original product life cycle theory, for example, only offered an explanation up to the stage of developing country firms exporting to AC markets: Stage 4 in his model.

Yun observed that, as relative latecomers to the global electronics industry, Korean firms had followed an internationalization process that differed from the sequence seen in industrialized countries. For the Koreans, the process began with import substitution and the development of the new industry through inward investment (Stages 2 and 3), and moved on to exports and production in advanced countries (Stages 4 and 5) before, as a final stage, undertaking production in and exports from LDC markets (Stages 6 and 7) (Yun 1993: 253–255, 263).

The decision to target industrialized markets first was a result of the policy of establishing mass production facilities, in order to achieve economies of scale and enhance price competitiveness. Only markets in industrialized countries had the capacity to absorb this level of output of electronics items. The ability to source low-cost parts and components and semi-finished goods from home had been a major factor underlying the decision to make investments in industrialized markets despite the lack of a firm-specific monopolistic advantage. Yun's observation of the need to sustain strong export performance to boost production at home and thus maintain economies of scale reflects comments made by Jun Yongwook[1] on the irreversibility of decisions taken by the parent company (Yun 1993: 255).

Figure 2.4 Yun Tong-jin's Post-Product Life Cycle Model (P*'osŭt'ŭ PLC Model*)

	AC (e.g. United States)	*LDAC (e.g. Japan)*	*DC (e.g. Korea)*	*LDC (e.g. Vietnam)*
Stage 1	Production of new product Exports focused on AC markets	Imports from AC	No demand	No demand
Stage 2	Investment in LDAC Investment in DC	Import substitution/ Production Exports to DC	Imports	Imports
Stage 3	Investment in DC Imports from LDAC	Exports to AC Investment in DC (local market sales)	AC investment in DC Import substitution	Imports
Stage 4	LDAC investment in AC Imports from DC	Investment in AC	Exports to AC	Imports
Stage 5	DC investment in AC	Investment in DC (Bypass exports)	Investment in AC	AC investment in DC Import substitution
Stage 6	Imports from LDC	Imports from DC	Investment in LDC	Exports to AC
Stage 7	Diversification of import sources	Diversification of import sources	Imports begin	Diversification of export markets

AC Advanced Country
LDAC Less-Developed Advanced Country
DC Developing Country
LDC Lesser Developing Country

Source: Yun Tong-jin 1993: 253.

An Se-yŏng's Competitive Advantage–Representative Specialization Zone Theory (*Taep'yo T'ŭkhwa Chidae Iron*)

Writing in 1995, An Se-yŏng offered a critique of FDI theories existing in the mid-1990s, before presenting his own contribution – the Competitive Advantage–Representative Specialization Zone Theory. In his analysis of the main western theories of FDI, An focused on product life cycle theory, monopolistic advantage theory, internalization theory, and the eclectic theory of international production. An noted that, while the product life cycle is 'one of the most important theories' for explaining FDI in technology-intensive sectors, it is unable to explain investment in simple labour-intensive industries.

Also, according to An, many MNEs from industrialized countries have adopted a strategy of marketing and selling new products simultaneously in a number of markets worldwide, rather than concentrating their sales in industrialized countries, as Vernon's theory would suggest. Furthermore, the fact that there have been many cases of FDI in technology-intensive sectors by firms from developing countries such as Korea since the late 1980s goes against one of the basic premises of the theory: that developing countries produce only standardized products (An 1995: 100–101).

In An's view, although Hymer's Monopolistic Advantage Theory can offer a satisfactory explanation of investment by industrialized countries such as the United States, which are 'at the top of the global technology ladder', it cannot offer an adequate explanation for FDI by less technologically advanced nations or developing countries. Although the theory explains the advantages that an MNE might possess *vis-à-vis* a local firm, it hardly touches on issues such as the locational advantages of the host country or the firm's motivation for investment. Most importantly, it is unable to explain the phenomenon of reverse direct investment by developing countries (An 1995: 122).

With regard to internalization theory, An argued that, given its assertion that firms engage in FDI to maintain their technological advantage, it cannot provide a satisfactory explanation of FDI in sectors such as textiles that utilize mature technology. Nor can it explain new forms of multinationalization that are emerging, such as technology-seeking investment by Asian firms. Finally, An criticized the eclectic theory of international production, principally on the grounds that it is unable to explain the phenomenon of companies engaging in FDI in the absence of location-specific advantages and with the principal aim of circumventing trade restrictions (An 1995: 134–139).

An's own theory centred on the concept of a 'representative zone of specialization' (*taep'yo t'ŭkhwa chidae),* which combines the competitive advantage possessed by a firm and the comparative advantage offered by a host country. The theory argues that the most appropriate location for production is a country in which the investing firm's competitive advantages match the host's comparative advantages. Over time, changes will occur in these competitive and comparative advantages due to movements in production factor prices (such as wage increases), shifts in government policy regarding the fostering of specific industries, foreign investment, imports and so forth, upgrading of the country's technological capability, changes in patterns of consumer demand due to rising income levels, fluctuations in the supply of industrial and technical manpower, and the availability and cost of land for industrial use (An 1995: 104–107). As a country's advantages change, firms which have been operating there will be forced to look overseas for a production location, where their own advantages can be combined with local comparative advantages to restore their international competitiveness.

An argued that the strength of his own theory lies in its ability to address simultaneously the questions 'Why do firms engage in FDI?' and 'Which firms from which countries invest where?'. However, An himself admitted that his theory cannot explain the key issue of reverse direct investment such as Korean investment in the United States (An 1995: 108).

In recent years a number of Korean students based in the United Kingdom have carried out studies of Korean investment in the European Union (EU), with specific reference to investment made by consumer electronics companies. Most recent among these are works by Byun Jae-woong (1995) and Park Gi-han (1996). In both cases, the writers made little if any reference to Korean sources and no mention of Korean contributions to the FDI theory debate.[2]

Noting the inadequacy of most existing theories to explain reverse investment, Byun presented five theoretical FDI models using Korean manufacturing investment in the EU as a case study. In presenting these models, Byun observed that the export-led development strategy adopted by successive governments in Korea and the country's resource endowments (shortage of natural resources, availability of labour and so forth) have had a significant impact on patterns of Korean FDI. Byun also highlighted the importance of location-specific factors in the decision to invest overseas (Byun 1995: 72–74).

In his 'economic development and trade-related dynamic FDI model', which is reminiscent of Dunning's Investment Development Cycle, Byun identified three stages in the development of FDI activity by a given country. In the first stage, firms active in light industrial manufacturing sectors engage in FDI in 'culturally familiar' locations. The levels of inward investment exceed outward investment at this stage, as foreign companies are encouraged to invest in the country, bringing with them much-needed technology and capital.

In the second stage, companies engage in market-oriented investment, and the phenomenon of reverse investment begins to appear as a response to trade friction and protectionism in export markets. At this stage, investment is defensive and undertaken on the basis of limited competitive advantages rather than clear monopolistic advantages. Outward investment now exceeds inward investment: this was the stage Korea had reached by the mid-1990s. In the final stage, investment is made in capital- and technology-oriented sectors such as semiconductors and automobiles, and is based on superior firm-specific advantages. In common with many other Korean scholars, Byun noted the defensive nature of Korean RDI and observed that it has been undertaken in order to circumvent trade restrictions (Byun 1995: 79–83).

In his thesis, Park Gi-han made some interesting observations about oligopolistic reaction theory. In particular, he noted the contrast between the dominant position enjoyed by Korean consumer electronics companies in their home market and their role as 'minor players' in global markets. (Jun Yongwook referred to this 'asymmetric market position' in domestic and global markets in his monograph on the oligopolistic behaviour of Korean consumer electronics firms [Jun 1988: 51].) This contrasts with oligopolistic firms in the United States; these companies tend to be major players both at home and overseas. Park noted that, whereas Knickerbocker had observed a tendency for US firms operating in a highly oligopolistic industry to refrain from excessive competition in the interests of maintaining equilibrium, Korean firms did not appear to fear the upsetting of the *status quo* by follow-the-leader patterns of investment, due to the relatively small impact of their investments on the global market (Park Gi-han 1996: 21–26).

MOTIVATIONS FOR INVESTMENT

Writing in 1984, Kim Kyu-ch'ang highlighted the differences between the motivations for FDI undertaken by firms from industrialized and developing countries. According to Kim, motivations for FDI that were specific to multinationals from developing countries included ethnic ties, political instability at home, and regional rather than global economies of scale (Kim Kyu-ch'ang 1984: 24–25). Pan Pyŏng-gil identified a number of motivations for Korean FDI including the need to penetrate and secure overseas markets, source raw materials, acquire technology, and seek greater production efficiency. However, he also noted that the most important strategic motivation was the defence of export markets in the face of trade friction and protectionism. Thus Pan highlighted at a relatively early stage the defensive nature of Korean direct investment in markets where barriers were being erected to restrict the entry of Korean manufactured goods (Pan 1985: 144–145 and 1996: 171).

Over the past twenty years, many Korean scholars have conducted empirical studies of Korean companies investing overseas, seeking to understand the motivations for this investment. The results of these surveys have shown strikingly similar results. One of the most recent analyses of motivations for Korean FDI was carried out by Nam and Slater, who identified a number of 'likely determinants' of Korean overseas investment. These were the opportunity to circumvent trade restrictions or consolidate and expand market presence, lower production costs, availability of raw materials, and access to technology and sources of finance (Nam and Slater 1997: 39–40).

CONCLUSIONS

A survey of the Korean literature has revealed a consensus among Korean scholars on the inability of western and Japanese theories to offer a satisfactory explanation for all forms of Korean FDI. Although it is agreed that Korean companies investing in developing countries do so on the basis of advantages in terms of appropriate technology, experience of operating in developing countries, ability to minimize production costs, and support from the home government, firms investing in industrialized markets either lack such an advantage, or possess one that is temporary in nature or too weak to offset the cost of foreignness.

During the early part of the 1980s, Korean companies engaged in reverse direct investment in western Europe and the United States, taking advantage of low production costs at home that could be internalized into semi-finished goods, parts and components for export to and assembly in the host country. However, changes in the internal and external economic environment in the late 1980s and early 1990s, most notably rising labour costs in Korea and measures to prevent the establishment of assembly plants in advanced countries, have eroded that advantage. Nevertheless, the early years of the 1990s saw a rapid increase in Korean global FDI, including a significant number of cases of reverse direct investment.

Reverse direct investment, undertaken in the absence of a clear firm-specific monopolistic advantage, has been characterized by Korean scholars as 'defensive', 'immature', and 'premature'. Jun Yongwook noted that Korean firms have engaged in RDI both as a means of protecting export markets and as a result of strategic decisions taken by the mother company, which could not easily be changed in the short term. This view was supported by Yun Tong-jin, who maintained that firms internalize not only monopolistic advantages but also changes in the external environment, such as the erection of trade barriers and changes in host government policy.

This survey of theories of foreign direct investment developed by western and Korean scholars has raised a number of issues and themes which will be addressed and developed further in the remaining chapters of this book. There are two major concepts from the Korean literature that have moved the debate forward and encapsulate these issues and themes. These are Yun Tong-jin's belief that external factors (that can be internalized) have led to Korean companies to engage in RDI, even though such investment was theoretically unfeasible or economically unviable ('involuntary internationalization') (Yun 1993: 242), and Jun Yongwook's belief that irreversible strategies adopted and policies implemented at home were a key factor in the decision to move overseas ('structural irreversibility') (Jun 1989: 182).

These concepts, together with the characterization of Korean RDI as premature and defensive, suggest that factors other than the possession of a firm-specific advantage have led to the rapid increase in Korean investment in the industrialized countries of western Europe since the mid-1980s.

These contributions to the debate on FDI in general and RDI in particular have opened up new areas for analysis and discussion. It is the author's belief that Korean direct investment is best understood from a theoretical/macroeconomic perspective, which can incorporate patterns of economic and industrial development, industrial structure, government policy, and changes in the domestic and global business environment.

If Korean consumer electronics firms have engaged in RDI on the basis of a temporary, weak or non-existent advantage, is it true to say that they were pushed or even forced to go overseas by changes in the domestic and global environment? Can it also be said that Korean government policy – economic, industrial and diplomatic – has been a factor in the rapid increase in Korean outward investment since the mid-1980s? Have the structure and characteristics of the consumer electronics industry made overseas production a matter of survival? Has the oligopolistic structure of the consumer electronics industry (dominated by three giant firms) led to follow-the-leader patterns of investment in Europe? Finally, have the location-specific advantages offered by Europe been a factor in investment decisions? It will be shown that, by answering these questions, a theoretical/macroeconomic framework can be created which will prove useful in explaining all forms of investment in Europe, from FDI in central and eastern Europe to RDI in western Europe.

Chapter Three

The Economic Context: The Republic of Korea (1948–1997)

Having examined the western and Korean perspectives on the theoretical foundation for foreign direct investment, the economic context in which foreign direct investment has occurred will now be analyzed. Of particular interest are the policies adopted by successive governments (which will be considered in depth at a later stage in the context of 'irreversible decisions') and changes in the domestic and global business environment that have had a significant impact on the Korean economy. These are changes that may have encouraged or even forced Korean firms to engage in overseas investment. The growth of the Korean conglomerates or *chaebŏl* will also be described, so that general patterns in their development may be understood and afford insight into the question of the impact of oligopolistic industrial structures within the Korean economy.

THE GROWTH AND DEVELOPMENT OF THE KOREAN ECONOMY

The five decades following the establishment of the Republic of Korea in 1948 saw the transformation of the South Korean economy from a small, stagnant, agricultural subsistence economy into one of the world's leading Newly Industrializing Economies (NIEs), along with Hong Kong, Singapore, and Taiwan. In 1996, the Republic of Korea became only the second Asian country (after Japan) to join the Organization for Economic Cooperation and Development (OECD). At the time of joining, South Korea was the 11th largest economic power and the 12th largest trading nation in the world (*Economic Report* November 1996: 26–27).

South Korea's success story can be summarized in a few remarkable statistics: between 1962 and 1997, the economy recorded an average annual growth rate in excess of 8.0 per cent. In little more than three decades, gross national product (GNP) expanded from US$2 billion to US$286 billion, *per capita* GNP rose from US$82 to US$9,511 (having peaked at US$10,543 in

1996), and exports soared from US$55 million to US$136 billion (at current market prices) (National Statistics Office 1997: 212–213, 223 and *Business Korea* October 1998: 82).

The nation's industrial structure has also undergone a transformation: the share of GNP accounted for by the manufacturing sector rose from 15.7 per cent in 1960 to 31.3 per cent in 1985, before falling back to 29.5 per cent in 1990. The share accounted for by the primary sector fell from 36.8 per cent to 9.0 per cent between 1960 and 1990, and the service sector's share of GNP, which stood at 47.5 per cent in 1960, had increased to 61.5 per cent by 1990 (Koo and Jo 1995: 49). In 1995, the respective shares of the primary and manufacturing sectors were 6.6 per cent and 26.9 per cent (Korea Foreign Trade Association 1994: 132).

As the economy has developed, there has been a switch in focus away from the production of labour-intensive, light industrial goods including textiles, toys, and footwear in the 1960s, to the development of heavy and chemical industries such as shipbuilding, petrochemicals, and automobiles in the 1970s. The 1980s and 1990s have seen a focus on technology-intensive industries such as industrial electronics and semiconductors, and the fostering of strategic industries for the 21st century, including aerospace, biotechnology, and advanced materials.

The Inter-War Years (1948–1953)

Prior to the establishment of two separate states on the Korean peninsula in 1948 (the Republic of Korea [ROK] in the south and the Democratic People's Republic of Korea [DPRK] in the north), the northern and southern halves of the peninsula complemented each other in terms of economic activity. The south was primarily agricultural, producing almost two-thirds of the country's food grains, and was the location for 80 per cent of Korea's light industrial facilities. In addition to housing 70 per cent of Korea's heavy industry, the north produced 90 per cent of the nation's electric power, almost 80 per cent of its mining output, and the majority of its chemical production (Sakong 1993: 2 and Nahm 1988: 479–480).

The economic chaos that followed the division of the country led to food and power shortages in the south, as agricultural production declined without the necessary fertilizers, and electricity supplies from the north were cut off. Factories closed due to shortages of power, materials, technicians and managers, and the unemployment rate rose. The newly-created Republic of Korea had virtually nothing to export and lacked the foreign exchange needed to import goods from overseas. Furthermore, with savings almost non-existent, there was no capital for investment in the development of new industries. Social problems exacerbated the economic chaos as Korea's rapidly expanding population was further boosted by the heavy influx of refugees from China, Manchuria, Japan, and North Korea (Nahm 1988: 351–354, 479–80).

One of the first actions taken by the Syngman Rhee government after it came to power in 1948 was the implementation of land reform, redistributing land bought from Korean landlords or confiscated from Japanese landholders at the end of the colonial period (1910–1945) and selling it to Korean farmers at less than market value. The government hoped that the former landowners would use the proceeds from the sale of their land to set up new businesses. Although the plan called for the payments to be made with securities redeemable over a five-year period, the real value of the securities fell due to hyper inflation and the government defaulted on some of the payments. In contrast, the Koreans who bought property formerly owned by the Japanese at heavily discounted prices went on to build enterprises that continue to play a key role in the South Korean economy today (See The *Chaebŏl*).

In the brief period between the establishment of the Republic in 1948 and the outbreak of war in 1950, the South Korean economy began to show some signs of improvement in terms of production. Between 1948 and 1950, industrial production increased by 50 per cent, coal production by 40 per cent and electric power generation by 33 per cent (Macdonald 1990: 195). However, the Korean War (1950–1953) destroyed what little industrial base had been created in the south. Total war damage was estimated at US$3 billion; two thirds of South Korea's production facilities were demolished, tens of thousands of buildings were destroyed, five million South Koreans were made homeless, and 1.5 million lost their lives (Nahm 1988: 482).

The Postwar Period (1953–1961)

According to Cho Soon, the 1950s are generally believed to have been a period of economic stagnation, during which economic planners focused their efforts on short-term import substitution rather than medium- to long-term growth (Cho 1994: 11–12). Nevertheless, there were some signs of economic recovery in the mid-1950s: between 1953 and 1957, GNP grew at a rate of 5 per cent per annum in real terms and by 1956, reconstruction was largely complete (Sakong 1996: 2). Light industries were established (including cement, food, textiles, and paper), transportation and communication systems were repaired, and the housing stock was rebuilt. Most of the investment was made possible with aid from the United States and international organizations, which amounted to US$3 billion by the end of the 1960s (Cho 1994: 11–14). Perhaps one of the most significant achievements of the period was the rapid expansion of educational facilities and the increase in student enrolment at all levels. This achievement laid the foundation for the rapid economic development seen in subsequent decades by creating a literate, well-educated workforce.

The 1960s

In 1960, the Syngman Rhee administration adopted a three-year economic development plan, but the plan was aborted when the president was forced to resign following large-scale demonstrations against his regime. The fall of the Rhee government was followed by a brief period of rule by the former opposition party, under the leadership of Chang Myŏn. The medium-range economic plan prepared by the new government was not implemented immediately due to the *coup d'état* in 1961, but it later became the basis for the first of Korea's five-year economic development plans (Sakong 1996: 48).

The economy that President Park Chung-hee inherited from his predecessors was still predominantly agrarian, with widespread unemployment, negligible savings, dwindling foreign aid, a limited domestic market, minimal exports, and only a small industrial base. Given that the country's sole advantage lay in its large, low-cost, well-educated labour force and the value that the workers could add to imported materials, Park adopted and implemented a policy of 'exports first' or 'nation building through exports' (*such'ul ipkuk*).

According to the new policy, economic development would be led by industrialization, which would, in turn, be driven by exports of mass-produced, light industrial goods. This would increase employment and national income, and provide a sound base from which to develop the economy further. Korean firms would be able to compete in world markets on the basis of price, exploiting their advantage of cheap labour costs in sectors such as plywood and textiles. These were areas in which technology was standardized and mature, and production costs were a crucial determinant of competitiveness (Cho 1994: 31).

Given Korea's low levels of domestic capital, the financial underpinning for the early plans came, of necessity, from foreign sources. Vital funds came from Japan, in the form of US$800 million in compensation (US$300 million in non-repayable grants-in-aid) as part of the 1965 treaty normalizing relations between Korea and her former colonial ruler. Over the following decade, aid, credit and loans from Japan exceeded US$1.5 billion, bilateral trade expanded to US$3.7 billion, and Japan became the biggest investor in Korea. The ROK government also secured funds through the country's support for the United States during the Vietnam War, in respect of goods supplied to the US military and projects awarded to Korean construction and transportation companies (Lone and McCormack 1993: 147–148). In addition to these sources of funding, the Korean government secured substantial amounts of foreign credit to finance its ambitious development plans.

A key feature of the new economic policy was the establishment of an export promotion system that offered incentives for the expansion of exports, including easy access to credit, tax favours, administrative support, the creation of industrial zones, wage controls, and favourable exchange rates. The government increased the supply of credit available for investment and tightly controlled its allocation, channelling low-cost funds into sectors and enterprises that it considered appropriate for development. Industrial estates where export firms could acquire land at favourable prices were created in the towns of Masan, Kumi, and Iri, and in every provincial capital. Also, exporters were given free rein in importing necessary raw and intermediate goods and components within the total amount of their export earnings. In addition, wage levels and union activity were strictly monitored and controlled by the government to keep competitiveness high and production consistent (Song 1994: 95–98). However, enterprises that failed to follow government guidelines and directives found themselves facing tax investigations with the threat of penalties and criminal prosecution.

Although the plan did result in the rapid expansion of exports and production, and an average annual GNP growth rate of 7.8 per cent (compared with a target of 7.1 per cent), levels of investment and savings fell below expectations (8.8 per cent and 17.0 per cent of GNP compared with targets of 9.2 per cent and 22.6 per cent) (Cho 1994: 29). Moreover, the easy access to credit for exporters, rapid expansion of money supply, and low levels of domestic savings led to high levels of corporate and foreign debt and an aggravation of inflationary pressures. Nevertheless, during the period of the first plan, the Park administration established a basic development pattern that was to be followed throughout the 1960s and into the 1970s.

However, the high-growth strategy that had proved successful in the 1960s was facing limitations by the end of the 1970s. Although exports were increasing at a rapid rate, there had been a huge increase in the trade deficit as imports expanded at an even faster rate. Korean manufacturers were struggling to maintain their export price competitiveness in labour-intensive, light industries in the face of rising wages at home and increasing competition from other developing and lesser developed countries. Despite a rapid increase in domestic savings, Korea was still heavily dependent on foreign funds for its investment finance, a factor that contributed to the nation's high levels of foreign debt.

The 1970s

In 1970, the government launched a stabilization programme that aimed to address the problems facing the Korean economy by slowing down the rate of expansion of domestic credit and imposing limits on foreign borrowing. The resulting fall in real growth rates, dampening of capital formation, and decline in exports prompted government action to reverse the stabilization effort and return to a more expansionist policy, for fear of losing the support of big business and the electorate (Haggard *et al* 1994: 25–30). Nevertheless, dissatisfaction with the Park regime grew, and, following the declaration of a state of emergency in December 1971, Park imposed martial law within Korea in October of the following year. The moves heralded the introduction of the *Yushin* (Revitalizing) Reforms, which consolidated the president's position while limiting political activity, all in the name of national security.

The earliest economic initiative of the Yushin era was the Heavy and Chemical Industry Plan (hereafter HCIP), which was announced in 1973, the year of the first oil shock. The plan, which was undertaken for both economic and political reasons, marked a significant shift in policy, moving away from the promotion of labour-intensive, light industrial production. The nation's economic planners were aware that, as wages increased, Korea's competitive advantage in light manufacturing industries *vis-à-vis* other developing countries would be eroded. The bureaucrats also recognized the need to enhance the nation's industrial structure and diversify the composition of its exports to ensure steady growth in the years to come (Haggard *et al* 1994: 33).

Developments in the international political arena also played a part in the decision to switch focus, most notably the *débâcle* of the Vietnam War and the announcement of the Nixon Doctrine in 1969, which called for greater devolution of security responsibilities in Asia and appeared to herald US withdrawal from the region. These events, together with the actual withdrawal of one US infantry division from the Korean peninsula in 1971, encouraged Park to promote investment in the heavy and chemical sectors in order to provide a foundation for building a strong, self-reliant defence industry (Sakong 1996: 42).

Industry-specific legislation was introduced, including the Petrochemicals Industry Promotion Act (1970) and the Steel Industry Promotion Act (1970); free trade zones were set up at Masan (1970) and Iri (1973), and petrochemicals and electronics complexes were constructed in Ulsan and Kumi. Special banking arrangements were made to offer priority loans to investors in heavy and chemical industry projects, and the government offered industry- and even firm-specific incentives (such as tax benefits, preferential interest rates, and policy loans) for strategic sectors including iron and steel, nonferrous metals, shipbuilding, petrochemicals, and electronic and electrical equipment (Cho 1994: 35–36).

Concerns about the wisdom of creating an energy-intensive industrial structure in a country that was completely dependent on imported fuel appeared to be well-founded when Korea suffered a serious blow from the first oil shock. As oil prices quadrupled, Korea's import bill increased, terms of trade deteriorated, and inflationary pressures mounted. Although the plan was abandoned in the interests of stabilizing the economy, it was revived in 1975 when the global economic situation improved (Chung 1990: 43).

In that year, Korea began to feel the benefits of the global recovery as well as the effects of the 1973 agreement among advanced nations on exchange rate adjustment. The agreement led to the appreciation of the yen and the German mark, which in turn enhanced the price competitiveness of Korean goods in world markets. Moreover, revenues from construction projects in the oil-producing nations of the Middle East added impetus to Korea's economic recovery (Cho 1994: 38–39). In 1975, the government formulated a strategy for penetrating the Middle East market and by the end of the decade, revenues from construction projects were offsetting two thirds of the nation's average annual oil payments. Exports recovered, the annual GNP growth rate passed the 8 per cent mark, inflation was brought under control, and the balance of payments situation improved (Haggard *et al* 1994: 42, 53).

Another important initiative of the mid-1970s was the establishment of the General Trading Companies (GTCs). In 1975, the government announced that a restricted number of large firms would receive administrative and financial privileges in return for acting as export 'windows' for small manufacturing companies (Haggard *et al* 1994: 53). The benefits were dependent upon achieving a minimum annual level of exports, which was raised at regular intervals (Song 1994: 99–100). The 13 newly-designated GTCs therefore moved swiftly to set up new companies which would help them meet their export targets, increase their respective group's export volume, and enable them to benefit from economies of scale (Cho 1994: 43–44).

Despite the recovery in economic performance in the mid-1970s, it was becoming clear by the end of the decade that serious imbalances had appeared in the nation's economic structure. The government's decision to channel credit to the developing heavy and chemical industries left companies in the light industrial sector struggling for funds. The HCIP had offered the conglomerates another opportunity for expansion, resulting in a greater concentration of economic power in their hands, to the detriment of small- and medium-sized companies. Gaps in income distribution between the urban and rural sectors were widening, leading to an exodus from the agricultural sector (Sakong 1996: 57–59).

By 1978, there were signs that the economy was overheating: demand for skilled workers in the expanding heavy, chemical and construction industries put general wage levels under severe upward pressure, while the rapid increase in corporate and personal income led to real estate speculation and increased demand for consumer durables. The consumer price index, which had averaged 11.6 per cent per annum between 1965 and 1973, soared to almost 25 per cent between 1974 and 1975. After falling back to 12.7 per cent in 1976–1977 the index rose again to reach 25 per cent in 1980–1981 (Amsden 1989: 101).

Rising labour costs in the late 1970s eroded Korea's international competitiveness at a time when export performance was deteriorating due to a slowdown in the world economy and increasing protectionism in global markets. Furthermore, the loss of competitive advantage in export-oriented light industries came at a time when, although investing heavily in the heavy and chemical industries, Korea had not yet established a competitive edge in those sectors. The rate of export growth fell every year between 1976 and 1979 and, in the final year of the decade, export volume declined for the first time in the nation's history (Sakong 1996: 65).

In early 1979, the head of the Economic Planning Board presented a major report on the economy to President Park, in which he tackled the controversial issue of the role played by the HCIP in the structural problems facing the economy. His arguments in favour of reforms, broad changes in the style of economic management, and the implementation of measures to liberalize and deregulate the economy were shared by other influential advisors. Although Park remained sceptical, he agreed to the implementation of the plan, and in April 1979 announced the enactment of the 'Comprehensive Measures for Economic Stabilization' (Haggard *et al* 1994: 62–64). However, he was assassinated eight months later by the head of the Korean Central Intelligence Agency and the country was once more plunged into political, social and economic turmoil.

The Early 1980s

The Chun Doo-hwan administration, which came to power in 1980, was faced both with short-term problems and the need to adjust economic policy in order to deal with long-term structural issues. Following the economic problems of the late 1970s, balanced economic growth, price stability, and market liberalization emerged as new policy issues. The new government moved swiftly to implement a number of measures to stabilize the economy, including a reduction in money supply, controls on wage increases, and proposals to boost productivity. The new administration also privatized the commercial banks and made some efforts to ease the controls on interest and exchange rates. As a result of the government's actions, the rate of inflation began to fall and the trade and current account deficits narrowed. (The improvement in the trade figures was also due to a fall in oil prices and a decline in interest rates.)

1982 was a turning point for the Korean economy as inflation dropped into single figures for the first time since 1965 (Amsden 1989: 101). Confident that inflation had been brought under control, the government took steps to boost the economy, including a relaxation of monetary policy, a depreciation of the wŏn, and an increase in investment in infrastructure (Sakong 1996: 69–70). Top priority was given to expanding exports to speed up growth, resulting in an overall improvement in economic performance. The trade deficit narrowed from US$4.4 billion in 1979 to US$1.8 billion in 1983, and the current account deficit shrank from US$4.2 billion to US$1.6

billion during the same period. The annual rate of GNP growth recovered from negative 3.7 per cent in 1980 to 5.9 per cent in 1981 and 12.6 per cent in 1983 (Korea Foreign Trade Association 1994: 131, 256–257).

In addition to the impact of these measures to stimulate the economy, Korea also benefitted from the effects of the 1985 Plaza Agreement, as the realignment of major currencies led to a further depreciation of the wŏn, enhancing the price competitiveness of Korean products in world markets. The trend toward lower inflation and smaller trade and current account deficits carried through into the late 1980s and, coupled with the 'Three Lows' (favourable exchange rates, stable interest rates, and stable or falling oil prices) paved the way for Korea's spectacular economic performance in the latter half of the 1980s.

The Boom Years (1986–1988)

In 1986, the Republic of Korea recorded the first significant trade and current account surpluses in its 38–year history. (Current account surpluses of a few million dollars were recorded in 1965 and 1977.) The following year, Korea recorded one of the highest GNP growth rates in the world (12.3 per cent), and exports of textiles and electronic goods both passed the US$10 billion mark. Total exports surged by 36.3 per cent year on year to reach US$46.2 billion, and the current and trade account surpluses swelled to US$9.9 billion and US$7.7 billion respectively (Korea Foreign Trade Association 1994: 131, 256–257).

1988 saw a third consecutive year of double-digit economic growth, bringing the average annual growth rate in Korea between 1983 and 1988 to more than 10 per cent. By the end of the year, the trade and current account surpluses had expanded to US$11.4 billion and US$14.2 billion respectively. Outstanding foreign debt had fallen to US$31.2 billion, and an increase in foreign assets had pushed the nation's net foreign debt down to US$7.3 billion, raising hopes that Korea would become a net creditor nation by the end of the decade (Korea Foreign Trade Association 1994: 131, 256–257, 270). However, even as the economy was moving into top gear, the government found itself facing protests from major trading partners concerned by Korea's success in their home markets.

The recording of a series of current account surpluses prompted industrialized nations to impose tighter import restrictions on Korean goods, and to demand that the ROK government allow greater access to its own market for foreign goods and services. Most vocal among the trading partners was the United States which, as Korea's largest export market, was running a deficit of more than US$10 billion with the Republic in 1987 (Korea Foreign Trade Association 1994: 212). To address the problem of containing the trade and current account surpluses and thus ward off trade friction with major trading partners, the ROK government increased the liberalization of imports, allowed an expansion in outward foreign direct investment by Korean firms, eased restrictions on foreign exchange transactions, and accelerated the repayment of outstanding foreign debts.

Despite Korea's stellar economic performance in 1986 and 1987, changes were occurring in the domestic and global environment that would lead to recession by the end of the decade. The implementation of democratic reforms in 1987 had been followed by the formation of effective trade unions for the first time in Korea's history. Collective bargaining led to high wage increases which were far in excess of increases in productivity gains, and which eroded the price competitiveness of Korean goods in overseas markets. The prolonged and sometimes violent labour disputes that accompanied the wage negotiations affected both Korea's production and exports, and prompted overseas buyers to look elsewhere in the region for a cheaper and more stable source of supply.

The rapid appreciation of the wŏn against major currencies further weakened the competitiveness of Korea's exports at a time when competition from other Newly Industrializing Economies and developing countries was intensifying. Korean exporters found themselves unable to compete on price at the lower end of the market, yet lacking the quality to challenge the Japanese at the top end of the market. At the same time, the liberalization of domestic markets brought a surge in imports and a new challenge to Korean manufacturers, this time on their own territory.

The effect of these factors was clearly demonstrated in the economic statistics for 1989 as the rate of economic growth almost halved, falling to 6.9 per cent in real terms. With exports edging up just 3.0 per cent year on year to US$61.4 billion and imports jumping by 17.9 per cent to US$56.8 billion, there was a sharp decline in the trade and current account surpluses to US$4.5 billion and US$5.0 billion respectively (Korea Foreign Trade Association 1994: 131, 256–257).

The message was clear: Korea had reached the end of that stage of its economic development where it could simply rely on high-volume exports of low-tech, low-priced, mass-produced goods. Korea's export performance had been too price-sensitive to weather the combined effect of currency appreciation and wage increases. To advance to the next stage of development, the Koreans had to improve their product quality and upgrade their technology in order to move into the production of high-tech, high value added goods.

Structural Readjustment (1989–1995)

In the late 1980s, the ROK government took action to boost domestic demand and allow it to compensate for the reduced contribution made to growth by the export sector. This would give exporters time to regain their competitiveness by upgrading technology and improving product quality.

The policy appeared to be successful as private consumption and domestic construction boomed, resulting in GNP growth rates of 9.6 per cent in 1990 and 9.1 per cent in 1991 (Korea Foreign Trade Association 1994: 131). However, consumer preference for imported goods led to a further increase in the trade deficit and a deterioration in the balance of payments situation, while the construction of two million housing units and a variety of social infrastructure projects led to shortages of manpower and materials, aggravating inflationary pressures. Within two years, the government was once again taking action to curb consumer spending and cool down the overheated construction market.

On the export front, the early 1990s saw an improvement in some of the factors that had an adverse impact on the economy in 1989. The steady depreciation of the wŏn helped exporters regain some of their lost competitiveness, and the spring wage negotiations were completed more quickly and smoothly than in previous years. Furthermore, export markets with long-term potential were opening up in the former communist bloc, which offered opportunities in terms of outward investment and joint ventures. The restoration of diplomatic relations with the former Soviet Union and the People's Republic of China in the early 1990s also opened up significant opportunities for Korean manufacturers and exporters in terms of investment and trade.

1991 saw a further expansion of imports, boosted by the increase in private consumption and the demand for capital goods needed in the process of restructuring. However, export growth rates remained sluggish despite the opening up of new markets in Latin America, South-east Asia, and the former communist bloc. The trade account went back into the red in 1990, with a deficit of US$2.0 billion, which swelled to a record US$6.9 billion the following year. The current account similarly returned to a deficit, recording red ink figures of US$2.2 billion and US$8.7 billion in 1990 and 1991 respectively (Korea Foreign Trade Association 1994: 131, 256–257).

Economic news for 1992 was mixed. To the Koreans, the overall picture was gloomy, with GNP expanding by just 5.0 per cent year on year. The slowdown reflected sluggish facility investment, negative growth in the construction sector, and a cooling-off in consumer spending. However, the slower demand for imports contributed to the shrinking of the trade deficit from US$6.9 billion in 1991 to US$2.1 billion in 1992, and the narrowing of the current account deficit from US$8.7 billion to US$4.5 billion. There were signs that inflation was being brought under control, with an annual increase of 6.2 per cent at the consumer level, down from 9.7 per cent in the previous year (Korea Foreign Trade Association 1994: 131, 256–257, 299).

When Kim Young-sam assumed the Korean presidency in 1993, the economy was on a downturn, with inflationary pressures increasing and exports sluggish. Despite the implementation of a '100–Day New Economic Plan' with sweeping measures to boost the economy, including a reduction in interest rates, a more flexible monetary policy, support for small- and medium-sized enterprises and facility investment, signs of an economic recovery failed to materialize by the spring of 1993 (*Business Korea* January 1993: 30–31).

The Five-Year New Economic Plan announced in July reflected a shift in focus and laid specific emphasis on the need to develop the initiative and creativity of the private sector and reduce the level of government control over the economy. The plan also incorporated a concept that was to become a key theme of the Kim administration: the concept of globalization (*segyehwa*), incorporating a long-term strategy to develop world-class business institutions and practices in Korea. Key features of the plan were the improvement of the environment for inward investment, the facilitation of outward direct investment to accelerate the globalization of Korean companies, and active participation in world trade negotiations (*Business Korea* October 1995: 13–15).

1993 proved to be a good year for the Korean economy, with the rate of economic growth expanding to 5.6 per cent. Exports rose by 9.6 per cent and facility investment increased as corporate confidence in the new administration grew. Korea recorded a trade surplus for the first time in four years, but concern was expressed that the recovery was due more to external factors such as the strength of the yen – which appreciated by 15.2 per cent against the dollar during the year (Kim Chong-tae 1993: 31–33) – the depreciation of the wŏn, and surging demand in the People's Republic of China for Korean products, than to the impact of government policy (*Business Korea* April 1994: 31–32 and *Economic Report* March 1994: 18–19).

In 1994, there were signs of a strong economic recovery, as the rate of growth accelerated to 8.3 per cent. Boosted by the continuing strength of the yen and surging demand in China and South-east Asia, exports showed their strongest expansion since 1987, rising by 17.1 per cent year on year to US$96.0 billion. However, with a 22.1 per cent jump in imports to US$102.3 billion, the trade and current accounts went back into the red (National Statistical Office 1997: 212, 233). Despite the improvement in economic performance, in the mid-1990s the Koreans found themselves facing a similar situation to that witnessed almost a decade earlier. Export performance was being boosted by favourable exchange rates, enabling Korean companies to compete on price rather than quality and raising fears of a decline in export performance once the yen began to fall back against the dollar (Kim Chong-tae 1994: 32–33 and Kiernan 1994a: 18–20).

At the end of 1994, some private research institutes were forecasting a deterioration of the business environment in 1995, with a stronger wŏn, rising raw materials prices, higher interest rates, and a slowdown in domestic facility investment, as major projects were completed (Kang 1994: 24–25). In the event, 1995 turned out to be another good year for the

Korean economy as Korea became the 32nd country in the world to achieve a *per capita* GNP level of more than US$10,000 (*Economic Report* May 1996: 10–11). The rate of economic growth edged up to 8.7 per cent, and exports expanded by 30.3 per cent to pass the US$100 billion mark – little more than three decades had passed since Korea broke through the US$100 million level. However, the trade and current account deficits also expanded from US$3.1 billion to US$4.7 billion, and from US$4.5 billion to US$8.9 billion respectively (National Statistics Office 1997: 212, 233).

1996 and 1997

1996 was the year in which the Korean bubble burst. The beneficial effects of the strong yen had begun to fade from the second half of 1995, and 1996 saw the erosion of Korean export price competitiveness by unfavourable exchange rate movements (*Economic Report* April 1996: 10–11). The economies of Korea's major export markets such as China, Japan, and the United States were showing signs of slowing down, and Korean corporations scaled back their plans for facility investment as export prospects became more uncertain (*Business Korea* May 1997: 15–17).

The statistics for 1996, the year in which Korea joined the OECD, were not encouraging. At US$23.7 billion, Korea's current account deficit was the second largest in the world, after the United States (Kim Chong-tae 1997i: 18). In the same year, the trade deficit jumped to US$4.7 billion to US$15.7 billion and the rate of economic growth fell from 8.7 per cent to 6.9 per cent (National Statistics Office 1997: 212, 233). Korea's total foreign debt soared to US$100 billion from US$43 billion four years previously, making the Republic the fourth largest debtor in the world (*Economic Report* January 1997: 16–17).

In early 1997, a string of corporate failures, including some of Korea's largest conglomerates, sent shockwaves through the economy. In July of that year, the South-east Asian currency crisis began with the fall of the Thai baht and, as the crisis spread to the rest of the region, foreign analysts and investors became increasingly uneasy about the soundness of the Korean economy and the health of its financial institutions. Following the decision by international credit agency Standard & Poor's to downgrade the Republic's long-term sovereign rating, foreign financial institutions began to turn down requests from Korean banks to roll over their loans, and demand for foreign currency to redeem foreign debts increased rapidly.

The drain on South Korea's foreign exchange reserves, as the central bank tried to meet domestic demands for foreign currency, and the financial authorities endeavoured to defend the value of the wŏn, brought the country to the brink of bankruptcy by the end of November. The Republic of Korea was in danger of defaulting on its debts, with usable foreign currency reserves standing at just US$7.3 billion (*Business Korea* October 1998: 54). In 1997, Korea's foreign debts were estimated at US$120 billion, with US$66 billion due for repayment by the end of the year (Kim Chong-tae 1997a: 18–21).

On November 21, the ROK government turned to the International Monetary Fund (IMF) for assistance as the country faced bankruptcy. On December 3, Korea and the IMF signed an agreement for a package of USS$58.3 billion in loans which would help the ROK government pay its foreign debts and reform the nation's financial sector and systems of corporate governance.[1]

THE *CHAEBŎL*

Korea's conglomerates (the *chaebŏl*) have been defined as

> Big business groups or conglomerates exclusively owned, managed and controlled by an entrepreneurial founder and his family (Lee Yeon-ho 1997: 18).

In the space of five decades, the *chaebŏl* have come to dominate the Korean economy, expanding their activities into almost every sector of the domestic market, and (in many cases) establishing a global presence through export sales and direct investment.

In 1996, total sales by the top 30 *chaebŏl* amounted to US$447 billion, with the top ten accounting for 83.2 per cent of the total, and the top five for 67.5 per cent. In terms of debt, the top ten conglomerates accounted for 75.2 per cent of the US$398 billion owed by the top 30 groups, while the top five accounted for 57.7 per cent of the total (Shin Sanŏp Kyŏngnyŏngwŏn 1998: 78) (See Table 1).

Table 3.1 The top 10 *Chaebŏl*, 1996

	Sales US$bn	*Assets US$bn*	*Debts US$bn*	*Equity Capital US$bn*	*Short-term Net Profit US$mm*
Samsung	84.1	92.6	76.2	16.4	237.4
Hyundai	82.7	69.8	57.3	12.5	147.1
LG	57.5	51.4	40.5	10.9	360.4
Daewoo	45.7	43.6	33.1	10.6	442.9
Sunkyong	31.6	28.3	22.5	5.8	308.6
Ssangyong	23.9	21.3	16.9	4.4	(148.8)
Kia	14.3	17.2	14.4	2.8	(141.0)
Hanhwa	11.8	16.6	14.6	2.1	(222.4)
Hanjin	11.8	20.8	17.8	2.9	(185.5)
Lotte	8.5	8.7	6.0	2.8	59.5

Source: Shin Sanŏp Kyŏngnyŏngwŏn 1998: 78.

The development of the *chaebŏl* has, to a great extent, mirrored that of the Korean economy as a whole. Although, in the past, the *chaebŏl* have been given much credit for their contribution to Korea's outstanding economic performance, their domination of the national economy has also laid them open to censure and led to calls for their reform.

The History of the *Chaebŏl*

Although the roots of the modern Korean corporation can be traced back to the end of the 19th century, few of the major *chaebŏl* active in Korea today can trace their history back before 1945 (Yu 1991: 40–41). Of the top 30 conglomerates active at the end of the 1980s, only six could trace their origins back before 1945. Sixteen had been established during the Rhee administration (1948–1960) and eight under the Park regime (1961–1979) (Kang 1996: 21–22). The short history of the Korean conglomerates is hardly surprising in the light of the country's early economic history. Although there was some activity by Korean entrepreneurs during the Japanese colonial period (1910–45), it was small in scale and focused primarily on consumer industries such as brewing and foodstuffs. It was only in the period after the Korean War (1950–1953), when much of the industrial base established by the Japanese had been destroyed, that the real growth of the *chaebŏl* began (Yu 1991: 40–41).

A number of important factors can be identified in the development and growth of the *chaebŏl* over the past five decades. These include the disposal of Japanese assets and the allocation of foreign aid between 1945 and 1960, the Park administration's export promotion policy and the sale of state-run firms in the 1960s, the development of the heavy and chemical industries, the establishment of the General Trading Companies (GTCs), and the Middle East construction boom in the 1970s.

After the Japanese withdrew from Korea in 1945, factories, railways, power stations, and communications networks amounting to 70–80 per cent of all the industrial assets located in the southern half of the peninsula were handed back to the Koreans. The Rhee administration sold off the property at discount prices and it was later alleged that seven of Korea's top ten *chaebŏl* families had used their political connections to win special favours and seal 'covert deals' (Lee Yeon-ho 1997: 22–23). However, it has also been argued that, although influence peddling could not be ruled out, the government's options were limited by the lack of entrepreneurs with any kind of track record, their own inexperience in administrative affairs, and a dearth of information (Kang 1996: 27).

A second important factor in the formation of the *chaebŏl* during the period following liberation from Japanese rule was the allocation of the aid supplied to Korea by the United States and the United Nations, enabling recipients to build industrial facilities financed by long-term, low-interest bank loans (Kang 1996: 28). In the aftermath of the Korean War, priority was given to the development of industries that could satisfy the nation's most basic needs of clothing, food and housing. With the advantage of low-cost funding, the *chaebŏl* became involved in these key areas of the economy and, as these industries were highly dependent on imported materials, they also became active in trading, which offered additional opportunities for growth (Yu 1991: 50–56).

Through its control over the allocation of capital and the approval of imported technology, the Park government was able to channel the nation's resources and the entrepreneurs' energies into areas it had targeted for development. The *chaebŏl* entered new business areas using loans and foreign technology allocated by the Park regime, enjoyed monopolies in their new sectors, and made handsome profits. This marked the beginning of a long and mutually beneficial relationship between the government and the *chaebŏl.*

Another factor in the *chaebŏl* growth was the sale of insolvent and state-run enterprises in the 1960s, which some have compared to the sale of Japanese property in the 1950s. Once again, the conglomerates received tax favours, seed money, special financing from the Bank of Korea, and generous repayment terms for purchases that greatly enhanced their asset portfolios (Yu 1991: 57–61).

During the 1970s, when the government was pursuing a policy of actively promoting the development of the heavy and chemical industries, the *chaebŏl* expanded considerably, diversifying into the new target sectors identified by the Park administration. The conglomerates moved swiftly into the six areas targeted for rapid development (steel, machinery, shipbuilding, oil refining, electronics and non-ferrous metals), making use of a golden opportunity in terms of government assistance, financial support, tax breaks, and the establishment of industrial complexes.

The need to import raw materials and natural resources to fuel the rapid development of the 1960s and 1970s in turn created a need to expand exports in order to pay for these purchases. The establishment of the general trading companies became another key factor in the expansion of the *chaebŏl*, with all but one of the GTC licences being awarded to conglomerates. The 1970s also saw the Middle East construction market enter a phase of strong growth, in which Korean firms were able to participate on the basis of experience gained working on construction projects during the Vietnam War. The *chaebŏl* quickly established or expanded their presence in construction and construction-related sectors, and used the oil dollars earned in the Middle East to expand further, investing in real estate and other speculative investments (Yu 1991: 73).

The close links between the government and the conglomerates, and the benefits enjoyed by the *chaebŏl* as a result of those connections, were a major contributory factor in the change in official and public attitudes toward the *chaebŏl* in the 1980s. Whereas in the previous two decades, the conglomerates were looked upon as a driving force in the development of the national economy, concerns were now raised about their close links with the government, their increasing size and expanding wealth, and the degree of influence that they exerted over the economy. As early as May 1974, the government had tried to take action to prevent the *chaebŏl* monopolizing the economy, by issuing special directives to force the conglomerates to list shares on the Korea Stock Exchange and thus open their empires to public scrutiny. However, few responded to the pressure and the government became less able to enforce the measures as the economy began to weaken in the mid-1970s (Lee Yeon-ho 1997: 27).

After taking power in 1980, the Chun government took measures to curb the power of the *chaebŏl*, including the 'Measures for the Reinforcement of Enterprise Competitiveness', which required the selection of core businesses, the sale of non-business real estate, and the disposal of unnecessary subsidiaries. The Act on Monopoly Regulation and Fair Trade (Fair Trade Act) was also promulgated with the aim of stimulating fair and liberal competition by regulating 'undue collaborative activities' and unfair trade practices (Lee Yeon-ho 1997: 47).

However, in the latter years of the Chun regime, government control over the *chaebŏl* weakened, as the strong economic performance of the late 1980s and mounting trade and current account surpluses encouraged the government to adopt a more generous attitude toward the conglomerates. President Chun himself announced that his government would no longer take steps to control or regulate the *chaebŏl*, but would allow them to focus all their efforts on exporting and thus boosting the nation's economic development (Lee Yeon-ho 1997: 59).

The 'economic euphoria' of the late 1980s and the recognition of the important role played by the *chaebŏl* in Korea's spectacular rates of export and economic growth carried on into the early years of the Roh T'ae-woo administration (1988–1992). Although restrictions on the conglomerates were not reinforced immediately after Roh took power, regulations were tightened towards the end of the decade, as economic performance began to falter and the popular mood once more shifted against the *chaebŏl*. Renewed accusations of government-business collusion, unjust accumulation of wealth, unfair competition among enterprises, and distortions in distribution of wealth and resources prompted the government to implement policies aimed at regulating the *chaebŏls'* influence over the Korean economy (Lee and Choi 1995: 95–98).

A recurring criticism of the *chaebŏl* has been that their growth has come at the expense of the small- and medium-sized enterprises (SMEs).[2] It has been argued that, given that the basic policy direction adopted by successive governments was to foster the conglomerates and so promote rapid economic growth, it was inevitable that SMEs suffered in terms of access to resources (including finance), high quality labour, and capital goods. Rising wages in the *chaebŏl*-dominated sectors pushed up wage levels for SMEs, leading to higher production costs; lack of finance undermined attempts to develop or acquire technology, and the gradual encroachment by big business into SME territory (such as the manufacture of parts and components) left smaller firms with a shrinking sphere of operations (Cho 1994: 64–65).

Since the latter half of the 1970s, successive governments have implemented measures aimed at assisting SMEs, including the establishment of the Industrial Bank of Korea, the Korea Federation of Small Businesses, the Small and Medium Industry Fund, and the Small and Medium Industry Promotion Corporation. Legislation has also been enacted with the aim of fostering the development of smaller firms, such as the Small and Medium Industry Startup Promotion Act and the Industrial Development Act (a short-lived attempt to rationalize Korea's unbalanced industrial structure). However, neither legislation nor affirmative action programmes have resolved the problems facing the SME sector, which remains weak to this day (Cho 1994: 76).

Although the government-business relationship appeared to be operating smoothly in the early 1990s, there was increasing antipathy among the *chaebŏl* chairmen toward the Roh regime. The opposition had gained control of the National Assembly, the mass media could be more critical of the government following the democratic reforms of 1987, and the *chaebŏl* became openly defiant as their economic and political power increased (Lee Yeon-ho 1997: 63).

As the Korean economy moved into recession at the start of the 1990s, anti-*chaebŏl* measures were tightened and the government stepped up its efforts to make the conglomerates specialize and restructure. Although this led to a deterioration in the relationship between the government and big business, the measures were successful in that the degree of economic concentration decreased and the financial status of the *chaebŏl* improved somewhat. The debt-to-equity ratio of the top 50 conglomerates fell from 524.0 per cent in 1980 to 410.6 per cent in 1987, and further to 368.9 per cent in 1991 (Lee and Choi 1995: 101–102).

Criticisms of the *chaebŏl* continued unabated, with the conglomerates being compared with an octopus, maintaining a powerful grip on all sectors of the economy, while smaller firms failed due to lack of funds. Critics claimed that, although the rapid growth of the *chaebŏl* was made possible by public sector subsidies, the founding families were reluctant to share the wealth and power they had acquired with the rest of society. Furthermore, the reliance of the corporate sector on bank loans meant that the *chaebŏls*' preferential access to policy loans and ordinary loans deprived smaller and weaker firms of vital finance (Sakong 1993: 185–186).

The inauguration of Kim Young-sam as President of Korea in 1993 heralded the beginning of a very different relationship between the government and big business. On assuming the presidency, Kim announced that he would not accept political contributions from business and declared an end to political bargaining in the economy. Once again, the *chaebŏl* were required to restructure their organizations; the Fair Trade Commission announced that the top ten groups would be required to select three core businesses and designate the companies that would specialize in those sectors, while the next twenty *chaebŏl* would have to choose two core business lines (Ungson *et al* 1997: 78–80). As a form of compensation, the government eased the credit restrictions on the *chaebŏl*, promising to release the top thirty groups from the requirement that they obtain prior bank approval for investment and real estate transactions by 1996 (Ursacki and Vertinsky 1994: 397).

The greatest shock in the history of the *chaebŏl* came in January 1997, when Hanbo Iron and Steel, the flagship company of Korea's 14th largest conglomerate, failed to honour matured promissory notes to the value of US$6.3 million. At the time of its collapse, Hanbo was estimated to have debts totalling US$6.6 billion (Kwon 1997: 18–19); the company had recorded losses of US$104.1 million in the first half of 1996, and its debt-to-equity ratio stood at more than 1,900 per cent in June 1996 (*Business Korea* February 1997: 12–13). The impact of the Hanbo bankruptcy was felt throughout the economy as banks were left with huge bad loans, and small- and medium-sized enterprises faced unpaid invoices for work subcontracted or ordered by the *chaebŏl*. Smaller firms also encountered increasing difficulty in obtaining finance, as banks and financial institutions were forced to tighten their loan policies (*Economic Report* March 1997: 19).

Hanbo's bankruptcy was followed by the collapse of Sammi Steel and the news that Jinro and Dainong, Korea's 19th and 34th largest conglomerates, were on the verge of bankruptcy. In July it was reported that the Kia Business Group had become the largest and perhaps the most famous conglomerate to be placed under the protection of its creditor bank to prevent a string of bankruptcies within the group (*Newsreview* 26 July 1997: 14–15). By November 1997, five of the nation's major conglomerates were being managed by their creditor banks; thousands of small businesses which had been dependent on these *chaebŏl* went bankrupt, and the unemployment rate was rising (Sohn Young-ju 1997: 22–23).

As the crisis unfolded in late 1997, it became clear that a significant amount of the blame for the nation's economic woes was being placed at the feet of the *chaebŏl*. Indeed, the conditions for the package of loans from the IMF included a requirement for improvements in the structure and governance of the giant Korean firms. Their financial structure was fragile and in urgent need of reform, with high debt-to-equity ratios resulting from their reliance on bank loans to finance their rapid expansion. As of the end of 1997, the average DER of the top thirty *chaebŏl* was 519 per cent, compared with averages of 154 per cent in the United States, 193 per cent in

Japan, and 86 per cent in Taiwan (*Business Korea* October 1998: 55–58). Aggressive expansion in the past had led to high levels of debt, but the *chaebŏl* had been able to overcome financial difficulties by channelling funds from good performers to poor performers within the group. However, this had become impossible when the recession hit the entire conglomerate (Kim Chong-tae 1997e: 16–17).

Under the IMF programme, the *chaebŏl* would be required to strengthen their financial structure by selling off businesses, raising money in the capital markets, or attracting foreign capital. However it was done, debt-to-equity levels were to be reduced to less than 200 per cent. The need for a major reorganization of Korea's corporate giants had never been more pressing nor more obvious, but only time would tell whether their commitment to reform would remain strong once the immediate crisis had passed.

The reforms would have a major impact, not only at home in terms of the sales, mergers and even inter-group swaps of *chaebŏl* subsidiaries, but also overseas, as Korean companies were forced to re-think their foreign investment strategies. Hyundai and Samsung announced that they would cut their overseas investments in 1998 by 30 per cent and would reduce their manufacturing presence overseas, particularly in areas such as western Europe where production costs are relatively high (Kim Chong-tae 1998m: 16–18).

CONCLUSIONS

The economic and industrial policies adopted by successive governments of the Republic of Korea have, generally speaking, placed their primary focus on rapid, quantitative growth and export-led industrialization. In the 1960s and 1970s, export promotion strategies provided incentives and rewards for companies that supported the government's development goals in general and its export promotion policy in particular. Financial and administrative assistance was focused on a small number of conglomerates, enabling them to expand rapidly into diverse industrial sectors, where they enjoyed monopolistic or oligopolistic positions, protected from domestic and foreign competition. Close links with the government and preferential access to resources provided a solid foundation from which the *chaebŏl* could expand rapidly.

In promoting the expansion of low-tech, labour-intensive, mass-produced goods for export, the government encouraged the conglomerates to capitalize on Korea's advantage of an abundant pool of low-cost labour, and achieve price competitiveness through economies of scale deriving from high-volume production. The success of this strategy led to rapid export growth, mainly on an Original Equipment Manufacture (OEM) basis, but also resulted in trade friction with major trading partners in the 1980s. Another negative effect of this policy was the lack of focus on the development of new technology, marketing skills and distribution networks.

Changes in the domestic business environment in the late 1980s and early 1990s, including rapid wages increases, high interest rates and soaring land prices, led to an erosion of Korea's price competitiveness and decline in export performance. For an export-led economy such as Korea's, a fall in exports is inevitably accompanied by a deterioration in overall economic performance. As part of the industrial restructuring that began in 1989, Korean companies moved overseas in search of lower production costs and the technology and skills they needed to upgrade their manufacturing capabilities.

An analysis of Korea's economic development also highlights the importance of changes in the global political and business environment. In the 1960s and 1970s, concerns about the threat posed by North Korea to the security of the Republic (both in military terms and as regards social unrest should the north's economic system be proved superior to that of the south) prompted both the desire for rapid growth and the adoption of the Heavy and Chemical Industry Plan. In the early 1990s, the end of the Cold War era and the establishment of diplomatic relations with former communist countries opened up significant new opportunities for trade and investment in Europe and Asia.

Korea's links with the global economy have served her well when favourable shifts in exchange rates, competitive global interest rates and low energy costs boosted the price competitiveness of Korea exports. However, Korea's heavy dependence on imported energy (principally crude oil) has left her vulnerable to rising prices, supply disruption and resource nationalism, just as the adoption of an export-dependent growth strategy has kept Korea hostage to the changing economic fortunes of major trading partners such as the United States and the western European nations.

In recent years, Korea's growing role as a member of the international business community has also led to new responsibilities. Membership of international bodies such as the World Trade Organization has restricted the ROK government's ability to direct the economy as it has done in the past, and heightened demands from trading partners for access to the Korean market for their own goods and services.

In this chapter, a number of important themes have been identified, namely the key role played by the government in determining the course and nature of economic development, the dominant position of the *chaebŏl* within the Korean economy, and the impact of changes in the global environment on economic performance. The following chapter will examine the development and characteristics of the consumer electronics industry, enabling us to incorporate these economic and industrial themes and characteristics into the theoretical/macroeconomic framework which will further our understanding of Korean foreign direct investment.

Chapter Four

The Development and Characteristics of the Korean Consumer Electronics Industry

The purpose of this chapter is to analyze the growth of the Korean consumer electronics industry, focusing on the sources of growth, the impact of ROK government policy, and the effect of changes in the domestic and global business environment. The characteristics of the industry will be examined, in order to determine whether they have fostered or worked against the acquisition of firm-specific advantages, enabled manufacturers to respond flexibly to changes in the business environment, and fostered the development of a balanced industrial structure.

Although much has been written on the development and characteristics of the Korean consumer electronics industry, most of the works that have appeared have been in the Korean language and have, therefore, remained inaccessible to many Western scholars. In 1987, Jun published an article on reverse investment by Korean consumer electronics firms and in the 1990s, two studies appeared by Suarez-Villa and Han (1990) and by Pecht *et al* (1997), but the focus of both these works was the industrial electronics sector. Bark's 1991 trade policy paper for the World Bank relating to trade friction, and the few studies published in the 1990s have tended to focus on technology, covering such areas as technological change (Bloom: 1992), technological innovation (Hobday: 1995), and technology transfer (United Nations: 1990). In addition, there are a handful of books that contain sections on industrial adjustment to trade restraints (Kim and Won: 1989), the growth and strategy of the consumer electronics industry and its comparative advantages (Yu Seong-jae: 1989).

In August 1998, Shin Sang Hyup published a monograph on Korean consumer electronics investment in the European Union, but devoted only a small section of the work to an analysis of the consumer electronics industry itself. Shin's focus was on the FDI activities of Korean firms, their globalization strategy, the impact of EU integration, the influence of host country incentives, and investment decisions. Korean language sources used by Shin relating to the consumer electronics industry were limited to annual statistics published by the Electronics Industries Association of Korea.

Shin's monograph was based on his doctoral thesis which was published in 1995; in the late 1990s three other theses were presented on the topic of Korean FDI in Europe, by Park Gi-han (Manchester Business School, 1996), Byun Jae-woong (University of Leicester, 1995) and Hong Sunghoon (University of Warwick, 1995).[1] Interestingly, none of these Korean nationals made much use of Korean sources in documenting the development of the Korean electronics industry, with references, once again, extending no further than the use of official statistics. Park offered the most detail of the three, but, even so, restricted his comments to a brief section covering the industry's development, an analysis of the industry's competitiveness and a structural analysis of global market share, product lines, technology and so forth, before presenting two case studies. Byun highlighted some of the problems facing the industry in his four-page overview of the consumer electronics sector.

Given the dearth of English language sources, the majority of the background information on the Korean consumer electronics industry presented in this section has been gathered from Korean language materials. Key sources of information about the industry were the Electronics Industries Association of Korea (EIAK *Han'guk Chŏnja Sanŏp Chinhŭnghoe*) and the Korea Institute for Economics and Technology (KIET *Sanŏp Yŏn'guwŏn*). KIET is a valuable source of industry-specific reports, such as those written by Kim Sŏk-hŭi *et al* (1989) on the effects of FDI on the Korean electronics industry, Pak Sŏng-t'aek *et al* (1993) on the FDI strategy of the consumer electronics industry, and Jun Yongwook (1988) on the oligopolistic behaviour of Korean firms in their internationalization process. (This work was published in English.) In the following year, Jun authored a monograph in Korean entitled '*Overseas investment strategy in the era of internationalization*', which was also published by KIET.

Additional background information on the internationalization process and FDI strategy of the Korean consumer electronics industry was gleaned from postgraduate dissertations and theses. Notable amongst these were the works written by Shim In-sŏp (1988), Pak Yong-hwan (1992), Ku Cha-yong (1992), and Chi Nam-ung (1994) on the internationalization of the Korean consumer electronics industry. Studies of FDI by individual consumer electronics companies included those by Lee Sŏn-sang (Samsung, 1989), Kim Chin-yŏng (Daewoo, 1990), Yun Tong-jin (Goldstar and Samsung, 1993), Yu Kwan-sŏn (Daewoo, 1994), and Ch'a Wŏn-dae (Samsung, 1997).

KOREA'S PLACE IN THE GLOBAL ELECTRONICS INDUSTRY

From its humble beginnings in 1959, the Korean electronics industry has grown into one of the Republic's leading export industries, and is widely recognized as having made a major contribution to the nation's economic development and growth (Suarez-Villa and Han 1990: 274). The past four decades have seen remarkable developments in the Korean electronics industry, progressing from the simple assembly of imported parts in the late 1950s through the development and manufacture of consumer electronics products in the 1960s and 1970s, and on to the production of advanced industrial electronics products, including semiconductors and telecommunications equipment in the 1980s and 1990s.

By the mid 1990s, Korea had become the largest manufacturer of colour television sets and microwave ovens in the world, taking global market shares of 40 per cent for these products. In addition, Korea was the largest producer of colour picture tubes and the second largest manufacturer of monitors in the world (*Economic Report* July 1994: 23). As can be seen in Table 4.1, Korea accounted for 4.9 per cent of global production of electronics products in 1996, with output totalling US$50.6 billion. In that year, the Republic ranked fourth in the world in terms of production, after the United States (US$298.5 billion or 28.9 per cent of the total), Japan (US$282.4 billion or 27.3 per cent) and Germany (US$52.4 billion or 5.1 per cent) (EIAK 1997b: 1).

The past fifteen years have seen a marked change in the structure of electronics production in Korea. In 1986, the parts and components sector accounted for the largest share of production (46.1 per cent of the total), followed by the consumer and industrial electronics sectors, with shares of 33.1 per cent and 20.8 per cent respectively. Ten years later, however, the balance had shifted, as parts and components had increased their share to 54.0 per cent, while industrial electronics had moved into second place (23.7

Table 4.1 Global Electronics Production, 1995–1996

		1995 US$bn	*1996 US$bn*	*% of Total Production*	*Growth Rate %*
	Global Production	**983.6**	**1,033.2**		**5.0**
1	United States	285.1	298.5	28.9	4.7
2	Japan	267.1	282.4	27.3	5.7
3	Germany	52.2	52.4	5.1	0.4
4	**Korea**	**49.4**	**50.6**	**4.9**	**2.4**
5	Singapore	39.8	42.7	4.1	7.2
6	United Kingdom	36.4	37.5	3.6	3.0
7	France	35.3	35.6	3.4	0.9
8	Taiwan	29.4	31.7	3.1	7.8
9	Malaysia	27.8	31.1	3.0	11.9
10	Italy	20.9	21.6	2.1	3.3

Source: Elsevier Advanced Technology: Year Book of World Electronics Data. (Reproduced in Electronics Industries Association of Korea 1997b: 1)

per cent) and consumer electronics had slipped to third position with 22.3 per cent of total output (EIAK 1997b: 16). In terms of global production, Korea's industrial electronics sector accounted for a 2.2 per cent share, while the consumer electronics and parts and components sectors took shares of 10.5 per cent and 9.2 per cent respectively (EIAK 1997b: 7).

In 1996, Korea ranked fourth in the world in the production of consumer electronics goods (after Japan, Malaysia, and the USA), third in the production of parts and components (after Japan and the USA), and ninth in the production of industrial electronics products (after the USA, Japan, Germany, the United Kingdom, Singapore, France, Taiwan and Italy) (EIAK 1997b: 6). The fact that the industrial electronics sector accounted for the largest share of the world market in that year (63 per cent) but only accounted for 28 per cent of production in Korea (compared with 71 per cent in USA and Germany, 74 per cent in the UK, 55 per cent in Japan, 68 per cent in Taiwan and 62 per cent in Singapore) reflects the emphasis that had been placed by the Koreans on the consumer electronics and parts and components sectors in previous decades (EIAK 1997b: 2).

In 1995, Korea ranked seventh in the world in terms of exports of electronics products, accounting for 5.2 per cent of global exports and recording the fastest export growth rate in the world (39.3 per cent: See Table 4.2). In 1995, exports of parts and components accounted for 64.8 per cent of Korea's electronics exports, with a 9.3 per cent share of global exports. Consumer electronics goods accounted for 14.7 per cent of Korean exports and 6.8 per cent of global exports, while industrial electronics products took a 20.5 per cent share of Korean exports and accounted for 2.0 per cent of global exports. In terms of imports, Korea ranked 13th in the world, with US$15.3 billion out of a global total of US$620.0 billion (a 2.5 per cent share). The top three importers were the USA (US$ 144.0 billion and 23.2 per cent), Germany (US$51.1 billion and 8.2 per cent), and Hong Kong (US$46.7 billion and 7.5 per cent) (EIAK 1997b: 8–10).

Table 4.2 Global Electronics Exports, 1995

	Export Value US$bn	*Year-on-Year Growth Rate 1994–1995 %*	*Share of Global Exports %*
Japan	120.7	13.3	18.0
USA	98.8	19.0	14.8
Singapore	62.5	28.8	9.3
Hong Kong	48.4	29.4	7.2
Germany	43.8	23.7	6.5
United Kingdom	37.9	30.2	5.7
Korea	34.7	39.3	5.2

Source: Electronics Industries Association of Korea 1997: 8–10.

A HISTORICAL OVERVIEW OF THE DEVELOPMENT OF THE KOREAN CONSUMER ELECTRONICS INDUSTRY

The development of the industry can be divided into five phases:

1959–1968	Import substitution and domestic production
1969–1974	'Passive' export orientation
1975–1979	'Active' export promotion
1980–1987	'Defensive' overseas production
1988–1996	'Aggressive' bulk overseas production

(Adapted from Chi 1994: 49)

Import substitution and domestic production (1959–1968)

In the mid- and late 1950s, as the Koreans struggled to rebuild their industrial structure after the Korean War, demand for electronics products within Korea was satisfied by imports. Lacking in technology and financial resources, the emerging Korean electronics companies imported parts and components from overseas for assembly in and export from the Republic. Thus, from an early stage, the industry depended heavily on foreign firms to help them establish a technological base from which to grow. Local production of electronics items began in 1959, when Goldstar started to produce transistor radios, using parts and components supplied by the Sony Corporation (Park Gi-han 1996: 171).

In 1961, the ROK government took steps to promote domestic production of electronics goods, by banning imports of radios and initiating a movement to increase the diffusion of radios in farming and fishing villages. These measures, together with a strong focus on import substitution, led to a remarkable increase in domestic production of radios, with 1.3 million sets being manufactured in 1968 (Wŏn 1986: 77–78). Although most of the electrical goods manufactured in Korea were produced for domestic consumption, some of the output was destined for overseas markets. The first exports of radios from Korea took place in 1962, with an order from Hong Kong valued at US$4,000 (Chi 1994: 49).

In 1966, Goldstar began assembling black and white television sets using parts and components imported from Hitachi in Japan (Chi 1994: 49) and, in the same year, the electronics industry was designated a strategic export industry and targeted for accelerated development (Pak *et al* 1994: 34). In support of this, the ROK government enacted the Foreign Capital Inducement Act (*Woeja Toippŏp*), aiming to encourage inward investment by foreign firms that could supply the technology and know-how needed by Korean companies. Encouraged by the government's action, emerging Korean consumer electronics firms began to take positive steps to secure investment from overseas, expand local production and accelerate import substitution (Chi 1994: 50).

In the 1960s, the most significant advantage that Korea could offer to foreign investors was its abundance of good quality, low-cost labour, which made the country an attractive base for cost-effective, labour-intensive assembly operations. Following the approval for the first inward investment project in 1965 (the establishment of KOMI Semiconductors), foreign electronics companies entered Korea and established a strong position in the local market. By the mid-1970s, more than half the workforce in the Korean electronics industry was employed at foreign or joint venture firms (Park Gi-han 1996: 174–175).

During this period, the electronics industry passed through the first three stages identified by Yun Tong-jin in his Post-Product Life Cycle Model. Following a period in which there was no demand, companies began to import foreign goods before establishing a domestic industry on the basis of technology and capital sourced from overseas. During this stage, there was little incentive for outward investment, as the industry was in the early stages of development and, reliant as they were on foreign technology, Korean firms had no advantages to exploit overseas. On the other hand, investors from industrialized countries such as Japan and the United States, were attracted to Korean by low labour costs and the potential of the local market.

'Passive' export orientation (1969–1974)

The 1960s and 1970s saw a strong focus on the expansion of simple assembly operations using low technology and key parts obtained from overseas sources or through inward investment. Despite the steady growth in domestic demand for consumer electronics products in the 1960s, local manufacturers quickly found their growth constrained by the small size of the Korean market, and it became clear that manufacturers would have to export to survive. Between 1969 and 1974, the average annual growth rate for exports of black and white televisions was more than 200 per cent. By 1974, 4.3 million radios, 600,000 black and white televisions, and 29,000 colour television sets were being exported from Korea every year (Wŏn 1986: 78–79).

In 1969, the ROK government enacted the Electronics Industry Promotion Law (*Chŏnja Kongŏp Chinhŭngbŏp*) with the aim of promoting exports of electronics products. The government promulgated an eight-year development programme providing for the establishment of industrial estates and offering tax breaks and direct subsidies to firms which were making investments to increase their exports, buy foreign technology or establish export operations overseas (Suarez-Villa and Han 1990: 275). Inward investment was encouraged, on condition that all output was exported and in 1973, Korea National, a joint venture formed by Anam Industries and National Electric of Japan, produced the first colour television sets in Korea. Despite the rapid increase in export activity during this period, Chi Nam-ung termed this phase of development 'passive export orientation', as export activities were hampered by Korean firms' lack of information, export experience, and overseas sales capabilities (Chi 1994: 50–51).

From the late 1960s, the ROK government took action to foster the rapid development of the electronics industry and, from this point on, the pressure grew for the newly-established electronics firms to expand production and achieve high levels of export sales. With the rapid increase in output and the relatively small scale of demand at home, manufacturers were forced to look to overseas markets to maintain the levels of production and exports that would ensure continuing government support and rewards.

Active export promotion (1975–1979)

From the mid-1970s, Korean firms increased their development and manufacture of new products; in 1976 electronics production and exports, which had passed the US$100 million mark in 1971 and 1972 respectively, broke through the US$1 billion mark (Kim Sŏk-hŭi *et al* 1989: 30–33), and production increased at an average annual rate of 47.2 per cent during the 1970s (Pak *et al* 1993: 34). Accounting for 13.4 per cent of total overseas shipments, the electronics industry was recognized as one of Korea's leading export industries and became the target of the government's preferential export promotion policy (Chi 1994: 51).

In the consumer electronics sector, production increased from US$270 million in 1975 to US$1.4 billion in 1979. During the same period, exports rose from US$199 million to US$925 million, accounting for more than half of all electronics exports from Korea. The late 1970s saw the development of two major products in the sector, namely the microwave oven in 1978 and the video cassette recorder (VCR) in 1979. Korea thus became the fourth country in the world to develop the VCR (following the USA, Japan and the Netherlands); Goldstar followed Samsung's achievement two years later, with Daewoo developing the product in 1982 (Park Gi-han 1996: 177).

In the latter half of the 1970s, Korean electronics firms made use of government support and drew on their own export experience to expand their overseas sales activities and penetrate new markets. In doing so, they were able to exploit their advantage of strong price competitiveness derived from economies of scale and low labour costs. (This is the fourth stage of Yun's Post-Product Life Cycle Theory.) From the mid-1970s, the larger Korean electronics manufacturers such as Samsung Electronics and Goldstar began to set up local marketing and sales companies in the United States as part of their export promotion policy.

However, the success of Korean consumer electronics products in overseas markets led to trade friction, prompting the ROK government to impose voluntary export restraints on shipments of colour televisions to the United States in 1978. Despite this action, later in the year, the US government imposed quotas on colour televisions from Korea and Taiwan through an Orderly Marketing Agreement (OMA). Quotas were based on the previous year's export volume: in Goldstar's case, the allocation was 91,000 sets, an amount equivalent to just 15 per cent of the firm's total capacity and representing 54 per cent of the export volume for the previous year. The OMA was renewed in 1979 and 1980 (Jun 1987: 97–102). The US government's action marked the beginning of a series of trade restrictions imposed on Korean consumer electronics products in America and Europe. This action constituted a major threat to export performance by the Korean consumer electronics manufacturers, who had focused their export activity in these markets (See Characteristics of the Korean Electronics Industry).

During this period, the ROK government took steps to promote the development of the domestic parts and components industry in order to increase the international competitiveness of Korean electronics goods and enhance the structure of the industry as a whole. The aim of this policy was to upgrade the production of consumer electronics goods from simple assembly of imported parts and components to technology-intensive, high value-added manufacture (Chi 1994: 52 and Wŏn 1986: 79). The local production of core parts and components would also reduce production costs and lessen Korea's dependence on overseas suppliers of these items.

The latter half of the 1970s saw Korean consumer electronics companies stepping up their efforts to secure and expand overseas markets for their products, supported, as always, by government policy. With the electronics industry now emerging as a leading export sector, efforts to boost overseas sales would make a major contribution to the nation's overall economic performance. However, the very success of this export drive began to draw adverse reactions from major trading partners, most notably the United States. What began as an agreement to limit Korean colour television exports to the US would spread to other markets and a wide range of products by the end of the next decade.

Defensive overseas production (1980–1987)

In the early stages of the development of the Korean electronics industry, the focus was very much on the consumer electronics sector; by 1982, after two decades of development, the industrial electronics sector still only accounted for 14 per cent of production (Wŏn 1986: 104–105). Between 1980 and 1987, the consumer electronics industry recorded an average annual growth rate of 29.4 per cent for production, 25.1 per cent for exports, and 16.2 per cent for domestic demand, which was boosted by the advent of local sales of colour televisions in 1980. Domestic production in the 1970s had been focused solely on export markets after the government banned local sales in order to curb excessive consumption and prevent 'social disharmony' (Yun 1993: 110). In the latter half of the 1980s, the special demand created by the 1986 Asian Games and 1988 Seoul Olympics combined with the boom in the domestic market to boost production and sales of consumer electronics products.

In 1982, the ROK government prepared a Long-term Plan for the Enhancement of the Electronics Industry (*Chŏnja Kongŏp Kodohwa Changgi Kyehoek*). Taking the view that the growth of the electronics industry would contribute to the attainment of high growth rates for the economy as a whole, the government reaffirmed its status as a strategic industry and established a fund to provide preferential financing for firms in this key sector (Chi 1994: 53). The plan called for an average annual export growth rate for electronics products of 18.6 per cent between 1982 and 1991, with Korea's global market share rising from 2.3 per cent in 1982 to 4.2 per cent by 1991. Greater emphasis was to be placed on the industrial electronics sector, which was to account for 31 per cent of total production by 1991, up from 14 per cent in 1982. The plan also aimed to enable Korean firms to break free from assembly production and strengthen their own product development capabilities. Financial support would come from the Electronics Industry Promotion Fund (*Chŏnja Kongŏp Chinhŭng Kigŭm*), which would provide funding on favourable terms for R&D and new product development. Direct financial assistance would be supplemented with other forms of policy financial support, such as export finance and tax breaks (Wŏn 1986: 104–105).

As Korean electronics firms gradually strengthened their presence and expanded their market share overseas, import restrictions against goods from Korea increased in key markets such as the United States and western Europe. By 1987 there were restrictions in the form of anti-dumping duties, quota restrictions, quality standard restraints and so forth on Korea's major export products (including colour televisions, microwave ovens, radios, and compact disc players) in these countries (Shim 1988: 42).

The first half of the 1980s saw a further strengthening of government support for the electronics industry, and the implementation of additional measures to boost Korea's ranking in the global industry. However, mounting concerns about the possibility of trade friction and protectionism in major export markets, coupled with the attraction of incentives offered for inward investment in those markets, were important factors in encouraging Korean consumer electronics firms to make direct investments in industrialized markets. The fifth stage of Yun's model sees consumer electronics companies establishing production facilities in industrialized countries. At these plants, low-cost parts, components and semi-finished goods imported from Korea were assembled, thus exploiting Korea's country-specific advantage of low production costs.

The erosion of Korea's export competitiveness

Before moving on to consider the final phase of growth, it will be useful to examine the reasons behind the decline in competitiveness that affected the export performance of Korean consumer electronics companies in the late 1980s. Under an export- and volume-led development strategy, Korean electronics firms planned for rapid growth through the mass production of a limited number of products (using imported technology, parts and components) and exporting them to a small number of industrialized markets.

Although the Koreans achieved high rates of growth in production and exports in the 1970s and early 1980s (See Appendices B, C and D), the latter part of the decade saw a sharp erosion of the price competitiveness on which Korean exporters (and, latterly, outward investors) had focused and relied. The domestic factors in this loss of competitiveness were rising wages, labour problems, and the strong wŏn. The external factors reflected, in part, a number of structural problems that had emerged as a result of the growth path followed by the industry.

Rapid wage increases

The advantage of abundant, low-cost labour that had bolstered the industry through to the late 1980s disappeared after 1989 due to a series of sharp wage increases, following the implementation of democratic reforms in 1987 and the subsequent creation of effective labour unions. Although hourly manufacturing wages in Korea still remained well below those earned in industrialized countries in 1990, they had risen at an average annual rate of 25.2 per cent between 1985 and 1990 (See Table 4.3). This rapid increase in labour costs was a key factor in the erosion of Korean industries' export price competitiveness in the late 1980s.

Despite further annual wage increases of as much as 56 per cent between 1988 and 1992, the stagnation of export demand due to saturation in industrialized markets and the emergence of competitors from less developed countries forced Korean manufacturers to lower their export prices, which led to even smaller profit margins. By 1992, 68 per cent of consumer electronics firms were operating at a loss or had an export margin of less than 2.0 per cent (Pak *et al* 1993: 43).

The loss of price competitiveness created an awareness of the need for Korean industry to move away from labour-intensive mass production to capital- and technology-intensive production. Moreover, rapid rises in royalties and increasing technology protectionism were forcing firms to

Table 4.3 Hourly Manufacturing Wage Rates, 1980–1990

	1980 $	*1985* $	*1987* $	*1988* $	*1989* $	*1990* $	*Growth Rate* *'80–'85* %	*'85–'90* %
Korea	1.03	1.35	1.78	2.50	3.57	4.16	5.6	25.2
Taiwan	1.00	1.50	2.26	2.82	3.53	3.98	8.4	21.6
Singapore	1.49	2.47	2.31	2.67	3.15	3.78	10.6	8.9
Japan	5.61	6.43	10.83	12.80	12.63	12.84	2.8	14.8
Mexico		1.60	1.06	1.32	1.60	1.85	N/A	2.9
USA	9.84	12.96	13.40	13.85	14.28	14.83	5.7	2.7
Germany	12.33	9.56	16.91	18.04	17.51	21.30	(5.0)	2.7
UK	7.43	6.19	8.97	10.46	10.43	12.42	(3.6)	14.9
France	8.94	7.52	12.42	12.96	12.60	15.25	(3.4)	15.2

Source: Ku Cha-yong 1992: 40.

consider investing time and money in developing their own technology. Investment in R&D would have to be accompanied by investment in capacity expansion and the automation of production processes in order to reduce costs and enhance product quality. However, operating on such thin margins, Korean firms had little spare capital to re-invest in their businesses.

The challenge from lesser developing country firms

Another factor in the erosion of Korea's international competitiveness was the increasing challenge from lesser developing country firms which were producing independently or on an OEM basis for Japanese manufacturers. In both cases, the firms were exploiting the advantage of a low-cost labour base that was now lost to Korea. As the Republic had done in the 1960s, these countries were taking positive steps to attract foreign investment, at a time when the Japanese were once again moving overseas to avoid the negative effects of the strong yen. The threat was strongest from Malaysia and Thailand in the labour-intensive, medium- to low-technology product markets; between 1985 and 1988, the Malaysian and Thai electronics industries grew at an average annual rate of more than 20 per cent (Pae *et al* 1990: 55).

Trade friction in industrialized markets

A third factor was the threat to economies of scale (and thus price competitiveness) arising from trade friction in industrialized markets. According to Pae *et al*, import restrictions since the 1970s have targeted labour-intensive, light industrial goods and consumer electronics products from Korea. Although, generally speaking, tariffs did not present a major obstacle to Korean electronics exports (after the Tokyo Round of 1973–1979, average tariff rates in industrialized countries fell by between 3 and 5 per cent), new obstacles appeared in the form of non-tariff barriers. This shift was signified by the introduction of Voluntary Export Restrictions, Orderly Market Agreements, and anti-dumping investigations in the United States and other key markets. In 1988, out of all Korean electronics exports to industrialized markets, almost 20 per cent faced trade restrictions of one kind or another, ranging from 14.2 per cent of total exports to Australia to 33.7 per cent in the EC (Pae *et al* 1990: 48).

Exchange rate fluctuations

Fourthly, the benefits deriving from the low value of wŏn against the Japanese and Taiwanese currencies between 1985 and 1987 weakened in the late 1980s as the wŏn began to strengthen, making Korean products more expensive in export markets. The wŏn appreciated by 3.3 per cent against the dollar in 1986, by 8.7 per cent in 1987, and by a further 15.8 per cent in 1988. With the exception of Taiwan, Korea's competitors saw their currencies depreciate against the dollar or remain stable; accordingly Korea's price competitiveness fell behind that of countries like Hong Kong, Japan, Singapore, and exports declined sharply. Although the rate of appreciation slowed after 1989 and there was some degree of exchange rate stability, the fluctuations had a major impact on the Koreans' export competitiveness and profitability (Pae *et al* 1990: 60).

Lack of technical manpower

Fifthly, at a time of urgent need for increasing numbers of skilled technical personnel to cope with the acceleration in the rate of technical innovation, Korean competitiveness was adversely affected by the lack of high-level technical manpower. Sŏnu *et al* note that Korean firms found the accumulation of technology difficult due to the lack of experienced R&D-related personnel and frequent changes in research topics, which made it difficult to maintain consistency in technology development. In the development of core parts and components, there were many cases of firms taking a short-term view and suspending or stopping their R&D activities when there was a drop in the price of items imported from Japan (Sŏnu *et al* 1990: 41).

High cost of financing

Finally, Korean firms faced high interest rates at home and restrictions on overseas borrowing in markets where rates were lower. In 1987 and 1988, financial costs for firms in the consumer electronics sector were on average three times higher than in Japan (Sŏnu *et al* 1990: 35–37).

'Aggressive' bulk overseas production (1988–1996)

1988 was a landmark year for the Korean consumer electronics industry in a number of respects; most notably, Korea became the second largest producer of consumer electronics goods in the world after Japan. However, in spite of this impressive world ranking, Korea's production capacity stood at no more than 20 per cent of that of Japan. In 1988, Korean consumer electronics production totalled US$6.1 billion, compared with US$31.7 billion for Japan and US$6.0 billion for the United States (Pak *et al* 1993: 34–36).

It was also in 1988 that electronics overtook textiles as Korea's leading export industry; in less than two decades (1970 to 1988), exports had risen from US$55 million to US$15.2 billion, and the share of total exports accounted for by electronics products had jumped from 6.6 per cent to 25.0 per cent (Pae *et al* 1990: 30–31). Finally, in 1988, production of sophisticated electronic components (including semiconductors) surpassed the production of consumer electronics for the first time, signalling a shift in the structure of the electronics industry in Korea (Suarez-Villa and Han 1990: 273).

In the late 1980s and early 1990s domestic demand for consumer electronics products remained strong due to rising disposable income, falling special excise taxes, the development of new products, and continuing restrictions on imported goods. However, the erosion of Korea's export price competitiveness, increased trade friction in and saturation of

Table 4.4 The Korean Electronics Industry, 1988–1993

	1988	*1990*	*1991*	*1992*	*1993*
GNP[a]	$172.8bn	US$251.8bn	US$292.0bn	US$305.7bn	US$328.7bn
E/E industry[b]	23.5	29.9	33.1	33.4	36.6
As % of GNP	13.6%	11.9%	11.3%	10.9%	11.1%
CE industry[c]	US$9.2b	US$10.1bn	US$11.1bn	US$10.5bn	US$11.2bn
As % of GNP	5.3%	4.0%	3.8%	3.4%	3.4%
As % of E/E	39.6	33.8	33.5	31.4	30.6
Exports	$60.7bn	US$65.0bn	US$71.9bn	US$76.6bn	US$82.2bn
E/E industry	15.7	17.8	20.1	21.6	24.2
As % of total	25.9%	27.4%	28.0%	28.2%	29.4%
CE industry	US$6.3b	US$5.7bn	US$6.1bn	US$6.0bn	US$6.5bn
As % of total	10.4%	8.8%	8.5%	7.8%	7.9%
As % of E/E	40.1	32.0	30.3	27.8	26.9

[a] GNP Gross National Product
[b] E/E Electronics and electrical goods
[c] CE Consumer electronics

Source: Yu Kwan-sŏn 1994: 11.

major markets, failure to develop new products to replace popular but mature items, and the strengthening of technology protectionism contributed to a deterioration in export performance. Exports of consumer electronics products declined in 1990 and 1992 before recovering to 1988 levels, and there was a significant drop in the share of total electronics exports accounted for by consumer electronics goods, from 40.1 per cent in 1988 to 26.9 per cent five years later (Yu 1994: 11).

In the latter half of the 1980s, Korean consumer electronics manufacturers began to switch focus to securing foreign markets and expanding overseas production. This change in strategy reflected a number of important developments, including the strengthening of protectionist measures in industrialized countries, the emergence of regional trading blocs such as North American Free Trade Agreement (NAFTA) and the European Union (EU), and the erosion of the competitiveness of Korean exports due to worsening production conditions at home. Other factors were the weakening of the benefits derived from the 'three lows' (low oil prices, favourable exchange rates, and low interest rates), and the challenge from lesser developing countries such as China and the ASEAN countries.

Reaching the sixth stage of Yun's model, Korean consumer electronics companies began to boost their investment in countries at a lower level of economic development, while retaining and even expanding their investments in industrialized countries. Although the competitive advantage once enjoyed in both trade and investment had weakened significantly, there was still a convincing argument to be made for maintaining a strong presence in the economic blocs that were emerging (See Chapter 5 The Korean View of European Integration). However, developing countries could offer lower production costs, good medium- to long-term market potential and, in some cases (such as the countries of eastern Europe), access to industrialized markets where trade restrictions were hampering export performance.

The ROK government's response to the slowdown in export performance in the early 1990s was to draw up a five-year plan to run from 1992 to 1996, which would help the electronics industry regain its competitiveness and ensure its continued growth. The focus of the plan was the development of core technology and the localization of key parts and components, thus reducing the industry's dependence on imported technology, avoiding technology protectionism, and reducing royalty payments.

The government identified three key areas for support: the development of 18 core technologies, parts and components such as liquid crystal displays which were highly capital- and technology-intensive, the commercialization of some 1,500 'bottleneck' technologies and test products, and the standardization of 200 parts and components already in production but not being sold among affiliated firms due to differences in technical standards. Preferential funding was offered for R&D activity and the construction of production facilities, and restrictions were placed on the imports of associated technology (Chi 1994: 57).

Despite corporate and government action, exports entered another slump in 1996, as the problems of high production costs, weakened competitiveness, adverse currency movements (a weaker yen and a stronger wŏn), falling product prices, and strong competition from less developed countries continued to affect Korea's export performance. In addition, the new offshore production plants were limiting direct exports from Korea, as demand formerly satisfied by exports could now be met by local production (Kim Chong-tae 1997j: 20). In 1996, although production increased by 15.5 per cent year on year to reach US$58.4 billion, exports declined by 5.6 per cent to US$41.2 billion (Kum 1997a: 33–34). Despite the weakness of its export sector, however, the electronics industry still accounted for 31.8 per cent of Korea's total exports and 19.6 per cent of total production in 1996. In the decade since 1985, the share of total exports accounted for by the electronics industry as a whole had risen from 15.8 per cent to 31.8 per cent, while the share of total output more than doubled from 9.2 per cent to 19.6 per cent (*Business Korea* October 1997: 35–37).

In the past decade, changes in the domestic and global business environment have had a significant impact on the consumer electronics industry. In common with other major industries in Korea, manufacturers of home appliances have faced rising production costs at home, increasing trade restrictions overseas, and fiercer competition in the domestic, as well as the global market. The search for lower production costs has prompted these companies to invest heavily in developing countries, while the need to avoid trade barriers and acquire technology and skills vital to industrial restructuring has attracted Korean investors to the industrialized countries of western Europe and the United States, despite the erosion of their manufacturing investment advantage in these markets.

THE CHARACTERISTICS OF THE KOREAN CONSUMER ELECTRONICS INDUSTRY

A survey of the Korean literature has revealed a number of characteristics of the Korean consumer electronics industry. Kim Sŏk-hŭi *et al* identified four key factors in the development of the electronics industry in South Korea: the availability of abundant, low-cost labour, active corporate facility investment, the inducement of foreign technology, and the ROK government's policy of fostering the industry and promoting inward direct investment (Kim Sŏk-hŭi *et al* 1989: 30–33). Other factors include a high degree of dependence on exports, a strong foreign presence in manufacturing and exports, lagging technology, a weak parts and components industry, an oligopolistic market structure, heavy reliance on a small number of markets, and the domination of production and exports by a few mature and standardized products. These factors and the development path followed by the industry have created a number of important characteristics which have helped to shape the FDI activities of Korean consumer electronics companies.

High export dependency

From the early stages of the industry's development, the limited size of the Korean market forced electronics firms to promote exports as a means of planning for the continued growth of the firm. Economies of scale (and therefore price competitiveness) could only be achieved through mass production, and the local market was too small and disposable income too low to allow them to rely on domestic demand to absorb their output. Government policy encouraged electronics firms to export by operating a system of incentives and penalties for exporting firms, and by imposing a ban on local sales of colour televisions in 1974, which forced manufacturers to find overseas markets for their products.

The export dependency ratio (the share of total sales accounted for by foreign markets) for the consumer electronics industry, which had stood at 51.9 per cent in 1970, gradually increased to reach 70.3 per cent in 1980, before falling back to 64.2 per cent by 1987. In that year the export ratios for the United States and Japan were 1.7 per cent and 43.1 per cent respectively. The import dependency ratio (the degree of dependence on imported parts, components, materials, intermediate and semi-finished goods) rose from 57.9 per cent in 1970 to 63.3 per cent in 1980, after which it eased back to 48.7 per cent in 1987. The import ratios for the United States and Japan in 1987 were 2.5 per cent and 8.3 per cent respectively. Given that parts and components accounted for a high proportion of imports, the statistics paint a picture of an industry in which materials, components, and intermediate goods were imported (mostly from Japan) for assembly and export (mostly to the United States and Europe). Even in the late 1980s and early 1990s, the export ratio stayed above the 50 per cent level, edging up from 54.8 per cent in 1991 to 56.6 per cent in 1992 (Park Gi-han 1996: 178).

This high level of export dependency has left Korea vulnerable to changes in the business environment. Over the past three decades, export performance has been affected by fluctuations in the economic performance of and consumer demand in key export markets, trade friction in the form of import restrictions, tariff and non-tariff barriers, the erosion of price competitiveness, and the emergence of competition from countries enjoying the advantage of low production costs.

Strong foreign presence in manufacturing and exports

Throughout its history, the Korean electronics industry has relied on assistance from foreign companies; in the early stages of the industry's development, the Koreans imported technology, parts and components to enable them to engage in assembly production of standardized goods. Lacking sales and distribution networks, they often relied on foreign partners to provide marketing, distribution and after-sales service through an Original Equipment Manufacture (OEM) agreement.

From the 1960s onward, the ROK government took positive steps to attract inward investment in the electronics industry; according to Kim Sŏk-hŭi *et al*, inward FDI made a larger contribution to growth in this sector than in any other industry in Korea. In cases where it was not possible to introduce the necessary technology directly from overseas (due to the reluctance of the owner to sell or license the technology), Korean firms had to acquire it through joint ventures and other forms of inward investment. Although the Korean government placed severe restrictions on inward investment up to the early 1980s, the Chun administration did implement a policy of positive inducement for investment in the electronics industry when it was accompanied by advanced technology. Encouragement was also given for investments located in an industrial complex (such as the one established at Kumi) or a free export zone, or investments that contributed to the expansion of exports. Although there were regulations controlling the amount of output to be exported, these did not prove to be a deterrent to investment (Kim Sŏk-hŭi *et al* 1989: 99).

Out of a total stock of US$700 million invested by foreign firms in Korea between 1966 and 1987, 33.8 per cent (US$239.9 million) was invested in the electronics industry. Out of a total of 336 investment projects undertaken in the electronics sector, 118 were initiated between 1966 and 1975, a further 84 between 1976 and 1985, and 134 between 1986 and 1987. Japan accounted for lion's share of the investment projects (215 or 64.0 per cent), followed by the United States (86 or 25.6 per cent). Almost half the investment projects undertaken (162 or 48.2 per cent) were in the parts and components sector, with industrial electronics a distant second (50 or 14.9 per cent) (Kim Sŏk-hŭi *et al* 1989: 55–58).

The boom periods in inward investment, which occurred between 1972 and 1974 and between 1985 and 1987, were due to a number of factors. In the early 1970s, there was a rush to invest by Japanese firms who were seeking low-cost labour and land, and were attracted by ROK government incentives to foreign investors. The strength of the yen created concerns over the possible loss of export price competitiveness and prompted Japanese firms to engage in FDI rather than exports. (This was a pattern that was to be repeated by Korean firms in the late 1980s.) By 1975, foreign firms and joint ventures accounted for 63.7 per cent of production and 73.9 per cent of exports in the Korean electronics industry, with the focus of investment on the parts and components and consumer electronics sectors (Kim Sŏk-hŭi *et al* 1989: 48–49, 59–61).

By 1980, 24.0 per cent of consumer electronics production and 29.0 per cent of exports was undertaken by foreign-owned firms or joint ventures.

The percentages were far higher for the industrial electronics and parts and components sectors (62.0 per cent of production and 67.0 per cent of exports in the case of industrial electronics, and 72.1 per cent of production and 76.0 per cent of exports for parts and components). Although these figures had improved by 1988, with less than 10 per cent of consumer electronics production and exports accounted for by foreign firms, the percentages were moderate for industrial electronics (around 28 per cent) and still high for parts and components (between 45 and 50 per cent), reflecting the weakness in these key sectors (Pae *et al* 1990: 36–37).

The characteristics of inward investment in the electronics sector include a preponderance of investment from Japan and the USA, which together accounted for 90 per cent of the projects undertaken and 80 per cent of the capital invested up to 1988. Most of the investment was made in the parts and components (including semiconductors) and industrial electronics sector, which together accounted for 71.4 per cent of inward investment projects in 1987. The past four decades have seen a shift in focus from semiconductor assembly in the 1960s to the manufacture of general parts and components and consumer electronics products in the mid-1970s, through to active investment in almost all areas (except consumer electronics) after the mid-1980s. The export ratio for foreign firms has always been high, given that one of the conditions for investment in the early days was the export of all output. However, the ratio has fallen steadily from 78.5 per cent in 1975 to 71.5 per cent in 1980 and 57.4 per cent in 1987, with a gradual switch in orientation away from exports and towards the domestic market (Kim Sŏk-hŭi *et al* 1989: 63–66).

Korean firms' dependence went further than the capital and technology acquired through inward investment; as noted above, they also relied on foreign firms for parts manufacture (through imports), marketing, distribution and after sales service (through OEM exports). Pae *et al* note that in the value-added chain for electronics firms (from research and development to parts and components manufacture, the assembly of finished goods, marketing, distribution and after-sales service) Korean firms focused on the assembly manufacture of semi-finished and finished goods, based on cheap and abundant labour and relied on foreign firms for other value-added activities (Pae *et al* 1990: 38). Thin profit margins from OEM sales made it difficult for firms to accumulate the capital needed for investment in R&D, brand name development, and marketing. In 1985, out of a total of 312 export items, fewer than half (43.3 per cent) were sold under Korean firms' own brands, reflecting the Koreans' continuing heavy dependence on foreign brand names (Kim Sŏk-hŭi *et al* 1989: 50).

For Korean firms, the main advantage of OEM exports was perceived to be the lack of necessity to invest time and money in the development of their own distribution networks and sales affiliates, as they could rely on the buyer and import agent to handle these activities. Mass production based on

OEM sales would reduce costs and the firm could avoid the burdens and risks associated with technology development. However, the negative effects of dependence on OEM sales later became apparent in the form of lower profits, lack of control over the market, fluctuations in order levels (making long-term production planning impossible), a lack of own brand image, and the danger that the buyer might switch to another supplier, become financially unstable, or even go bankrupt.

As late as 1989 the dependency rates for OEM exports were 53 per cent for colour televisions (the rate had peaked at 59 per cent in 1987), 57 per cent for refrigerators (79 per cent in 1986), 56 per cent for microwave ovens (60 per cent in 1987), and 22 per cent for video cassette recorders and washing machines (60 per cent in 1987 and 59 per cent in 1988) (Sŏnu *et al* 1990: 40). By the mid 1980s, Korean firms were realizing the limitations of OEM exports and were becoming more active in setting up their own distribution networks and after-sales service facilities in key markets, as well as making more effort to develop their own brand image and become directly involved in marketing their products (Pae *et al* 1990: 41–42).

The positive impact of a strong foreign presence in the Korean consumer electronics industry cannot be denied; in the 1960s and 1970s, inward investors brought with them much-needed capital and technology. This created the foundation for the industry to grow and expand, as Korean manufacturers acquired standardized technology suited to labour-intensive assembly and mass production. However, it can be argued that the focus on and success of OEM sales diverted the Koreans' attention from the need to acquire their own technology, develop and differentiate their own products and brand names, and create their own marketing and distribution networks.

Lagging technology

In the early stages of the industry's development, Korean firms focused on copying products already developed by firms from industrialized countries, and so paid scant attention to research and development. R&D investment levels were low, rising from 1.2 per cent of sales in 1976 to just 3.0 per cent by 1983 (Wŏn 1986: 99–100). This strategy of acquiring technology from foreign firms aimed to reduce the financial burden and time delay associated with in-house technology development. Although the Koreans did become active in technology development as the electronics industry grew, the Korean consumer electronics companies still lagged behind their competitors in industrialized countries in terms of the technology required for product design and innovation, the production of core parts and materials, and other key areas of activity (Park Gi-han 1996: 189).

According to a report published by the Electronic Industries Association of Korea, there was a three- to five-year time lag between Korea and industrialized nations in developing new products, and a five- to nine-year time lag in developing future generations of high tech products. The same report observed that Korean investment in R&D fell far short of the levels seen in Japan, with a ratio of R&D investment to turnover of just between 4 and 8 per cent in Korea against a ratio of up to 16 per cent in Japan (Park Gi-han 1996: 184–189).

From the above, it is clear that Korean consumer electronics companies have been unable to establish an industry- or firm-specific advantage in terms of superior technology or product innovation *vis-à-vis* their rivals in industrialized countries. The focus on quantitative expansion and rapid growth has, in the past, led to a concentration of effort on the acquisition of mature technology that can support the mass production of low-tech goods. While this technology may be exploited through investment in countries at a lower stage of technological development than Korea, the long-term future of the industry depends on access to advanced technology. The need for industrial restructuring that became clear in many of Korea's key industries in the late 1980s and early 1990s prompted some electronics companies to invest overseas in an effort to acquire vital technology either through collaboration with or the acquisition of technology-rich foreign companies.

Weak parts and components industry

In the 1960s and 1970s, the emphasis placed on the quantitative expansion of simple assembly operations in Korea using imported, standardized technology resulted in a lack of investment in the domestic parts and components industry. As we have seen in Chapter 3.1, the government's focus on the development of the *chaebŏl* in terms of access to capital, technology and manpower, stunted the development of small- and medium-sized enterprises. This was a contributory factor in the weakness of the parts and components industry, which could otherwise have been a key area for SME activity. The electronics industry as a whole has been characterized by high export and import dependency ratios and low rates for self sufficiency for core parts and components. In 1987, parts and components accounted for 75 per cent of total electronics imports, and exports of consumer electronics goods, parts and components accounted for roughly 80 per cent of total overseas shipments of electronics products (Kim Sŏk-hŭi *et al* 1989: 41–42).

The weakness of the parts and components sector and the heavy dependence on imports left Korean firms vulnerable to delays in supply, and adjustments to the volume and price of the shipments. Given that most technology came from Japan, the Japanese were able to force the purchase of core parts and components when supplying technology and then exercise control over the timing and volume of supply of the parts and components. Moreover, the crucial effect of the high royalties charged was an erosion of the price competitiveness of Korean goods. In the case of video cassette recorders and colour televisions, in the late 1980s royalties accounted for 8 per cent and 3 per cent respectively of the unit cost (Sŏnu *et al* 1990: 33–34).

The weakness of the parts and components sector argues, once again, against the possession of a clear competitive advantage in the Korean electronics industry. Heavy reliance on overseas sources for core parts and materials has left companies vulnerable to fluctuations in exchange rates, disruptions in supply, and externally-imposed increases in production costs.

Oligopolistic market structure

As has been discussed in Chapter 3, the *chaebŏl* have been a dominant force in the Korean economy, and the electronics sector offers a prime example of the oligopolistic market structure that developed in a number of industries. Until the end of 1998, the electronics industry in Korea was dominated by the 'Big Three', namely LG Electronics (formerly Goldstar), Samsung Electronics and Daewoo Electronics.[2] Such was the power of the 'Big Three' that Kim *et al (*1989) observed that the growth of the whole electronics industry depended on these three firms (Kim Sŏk-hŭi *et al* 1989: 52). In the early stages of the industry's development, Goldstar was the leading manufacturer and exporter of consumer electronics products in South Korea. However, the firm's dominant position was challenged with the emergence of Samsung Electronics Co. Ltd in 1969, and the establishment of Daewoo Electronics Co. Ltd in 1971. In 1984, Hyundai Electronics Inc. entered the market, but the company has placed more emphasis on the industrial electronics sector than its major rivals. By the mid-1990s, a number of second tier firms had emerged including Anam Industrial and Haitai Electronics, plus more than one thousand small- and medium-sized enterprises.

The ROK government's policy of offering support to a small number of firms with the potential to grow into world-class exporters and promoting volume-led industrial development in the 1960s and 1970s was a major factor in the creation of an oligopolistic market structure within the electronics industry. In addition to benefitting from the government's policies, the 'Big Three', as *chaebŏl* affiliates, enjoyed certain advantages over smaller domestic competitors in terms of their access to funding, their ability to establish and operate global production facilities, the support they received from other companies within the group and so forth. Within the *chaebŏl*, the electronics affiliates have been seen as flagship companies which make a major contribution to the groups' sales (both domestic and overseas), their profits, and their corporate image (Park Gi-han 1996: 192–196).

In 1986, Samsung, Goldstar and Daewoo accounted for almost two thirds (65.9 per cent) of electronics production in Korea; within the consumer

Table 4.5 The Big Three's Share of Consumer Electronics Production by Item, 1988

	1st	*Share %*	*2nd*	*Share %*	*3rd*	*Share %*	*Total %*
CTV	SS	33.0	GS	31.7	DW	17.5	82.2
BWTV	GS	32.2	SS	23.9	DW	8.7	64.8
VCR	SS	46.9	GS	38.3	DW	14.8	100.0
MWO	SS	42.7	GS	40.0	DW	17.3	100.0
RFR	GS	47.6	SS	36.7	DW	15.7	100.0
WM	GS	48.7	SS	35.1	DW	16.2	100.0

SS Samsung Electronics
GS Goldstar/LG
DW Daewoo Electronics

CTV Colour televisions
BWTV Black & white televisions
VCR Video cassette recorders
MWO Microwave ovens
RFR Refrigerators
WM Washing machines

Source: Pae Chong-gi *et al* 1990: 33.

electronics sector (excluding sound equipment) they accounted for about 90 per cent. By product, the 'Big Three' controlled the market for video cassette recorders, microwave ovens, refrigerators, and washing machines, and took shares of 64.8 per cent and 82.2 per cent of the production of black and white and colour television sets (Pae *et al* 1990: 33).

The intense competition among the 'Big Three' can be attributed to the need to expand exports to meet government targets and benefit from official incentives in the past and, in recent years, to the lucrativeness of the domestic market. Given this intense competition, it can be surmised that the three companies would respond promptly to new initiatives by a rival, either at home or in foreign markets, to maintain the *status quo* at home. As has been noted above, both western and Korean scholars have commented on the fact that the fear created by a rival's move into a new market will result in a loss of actual or potential market share for those who remain at home. Additionally, in the past, the prestige deriving from an investment overseas (in terms of international 'recognition' of the firm's products) and desire for government approval have acted as powerful incentives for following the leader into the new market (Park Gi-han 1996: 318).

Heavy reliance on a small number of markets

Just as they have relied heavily on the United States and Japan for capital and technology, Korean electronics firms have depended on the Japanese and American markets to source their imports and place their exports. In 1987, more than 50 per cent of exports went to the United States and Japan and more than 80 per cent of imports came from there. However, the degree of dependence was greater on the United States for exports and on Japan for imports, leading to a trade surplus with the United States and a trade deficit with Japan in the mid- and late-1980s (Kim Sŏk-hŭi *et al* 1989: 42–47). From the 1980s, Korean manufacturers began to target western Europe as a potential market for their consumer electronics products, in line with the ROK government's desire to reduce the country's dependence on these two markets (See Chapter 6).

The heavy dependence on a small number of industrialized countries reflects the fact that global demand for electronics products was concentrated in those markets. In 1995, the United States, Japan and Europe together accounted for more than 80 per cent of total world demand for consumer electronics products (Ku 1992: 28). Although the United States was the largest single market for Korean consumer electronics products, the Koreans' market share fell from 12.5 per cent in 1988 to 9.9 per cent in 1992. Japan's share also fell from 55.1 per cent to 46.1 per cent, but Malaysia increased its share from 2.0 per cent in 1990 to 10.7 per cent in 1992, to take the third largest market share after Japan and Korea (Pak *et al* 1993: 37–38). This reflects the erosion of Korea's competitive advantage after 1987; by the early 1990s, Korea had lost her advantage in terms of low-cost labour whereas companies in Malaysia still enjoyed that advantage and could exploit it in their export pricing.

Although the strategy adopted by Korean electronics manufacturers of flooding a small number of industrialized markets with their highly price competitive products (*chiptchung houshik such'ul*) was successful in the 1970s and early 1980s, it left the firms vulnerable to changes in the global trade environment (Pae *et al* 1990: 38). As has been noted above, the advent of a recession in a key market has, in the past, resulted in a sharp decline in Korean exports. American and European reaction to the inflow of low-priced Korean imports was swift and tough, with the imposition of trade barriers further threatening sales in Korea's largest markets. The decline in export performance witnessed in the United States and Europe at the end of the 1980s prompted many Korean companies to focus their efforts on diversifying their export markets. Fortunately, this coincided with the opening up of diplomatic relations with former communist countries in Asia, central and eastern Europe.

Dominance of a few mature and standardized products

As Korean electronics firms relied on imported technology to build the foundation of their industry, they focused on the acquisition of technology, parts and components for goods that had already passed through the development and growth phases of the product life cycle. Within those parameters, the large firms focused on a small range of products that could be mass produced, thus achieving economies of scale and enhancing export price competitiveness. In 1981, radio cassettes, video cassette recorders and colour televisions accounted for 64.1 per cent of production in the consumer electronics sector. This share fell to 63.6 per cent in 1983 and 54.2 per cent in 1985, before rising again to reach 61.1 per cent in 1988 (Pae *et al* 1990: 44). By 1991, audio products, colour televisions, video cassette recorders, and microwave ovens accounted for more than three quarters of total production of consumer electronics goods. The same four products accounted for 91.5 per cent of all consumer electronics products exports (Ku 1992: 39).

Just as too heavy a focus on a small number of markets brought problems for the Korean consumer electronics companies, so the concentration on production of a limited range of mature goods exported at low prices made it all the more likely that some form of trade barrier would be imposed, should the influx of Korean goods threaten the interests of the importing country.

CONCLUSIONS

In the 1950s and 1960s, Korea's factor endowments (principally a lack of natural resources and an abundance of high quality, low-cost labour) prompted the ROK government to adopt an export-oriented development strategy, in which emphasis was placed on quantitative rather than qualitative expansion. The government implemented a variety of policies aimed at boosting exports and building up strategic industries with the assistance of a small number of favoured business groups. Companies were offered incentives to invest in target sectors, including electronics, and were encouraged to build large-scale production facilities, achieve economies of scale, and thus enhance their export competitiveness.

The lack of indigenous technology, low levels of domestic savings, and the desire for rapid development forced Korean companies wishing to build up a presence in the electronics industry to rely on imported technology, capital, materials, parts and components. Imports and inward investment provided the means for the Koreans to build the foundations of an electronics industry, beginning with the simple assembly of imported parts and components. Given the existence of strong political and economic ties with Japan and the United States, Korean electronics firms came to rely on these countries, not only for the capital and technology they lacked, but also for research, marketing, and distribution functions.

The lack of focus on developing these vital skills and the concentration on price competitiveness left the Koreans vulnerable to changes in the business environment. The price advantage enjoyed by Korean exporters and investors up to the end of the 1980s was easily eroded by factors beyond their control, leaving them scrambling to establish low-cost production bases in developing countries, and forcing them to maintain and expand their existing investments in industrialized countries in order to secure the market and upgrade their technological capabilities.

The government's policy of helping a few large-scale firms with growth potential to build export industries created an oligopolistic market structure in the electronics industry, as in other sectors. These few firms focused on the mass production of a small number of products; with a domestic market too small to absorb their growing output, Korean electronics manufacturers looked to place their products in the United States and western Europe. The focus on these trading partners reflected the reality that theirs were the only countries with markets large enough to absorb Korea's export volume. Inevitably, though, Korea's export success led to import restrictions that threatened the companies' very survival.

An analysis of the development and characteristics of Korean electronics industry has revealed the crucial role played by the government, in terms of promoting inward investment (when deemed necessary), implementing measures to foster the development of the industry, and encouraging the creation of an oligopolistic industrial structure. The development path followed by the industry led to a heavy reliance on foreign technology, skills, parts and components, while the excessive dependence on a small number of markets to absorb huge amounts of output left Korean exporters vulnerable to changes in the business environment. Looked at from this perspective, the adoption of a high-growth policy made exports and, ultimately, foreign direct investment inevitable and even essential for survival.

Chapter Five

Korean Foreign Direct Investment

More than three decades have passed since Korean companies became active in foreign direct investment (FDI); nevertheless, in 1997 one Korean writer observed that Korean FDI has 'really only a ten-year history' (Lee Chae-min 1997: 4). Data published by Bank of Korea and shown in Table 5.1 reveal the truth of that surprising statement: as of 31 December 1996, Korean companies had invested US$13.8 billion dollars in 6,653 projects, on a net cumulative invested basis.[1] However, of that total amount, more than 95 per cent (US$13.1 billion) had been invested since 1987, and more than 90 per cent of the projects operational at the end of 1996 had been established within that same ten-year period (Bank of Korea 1997: 3). Between 1986 and 1990, overseas direct investment by Korean companies expanded at an average annual rate of 74 per cent (in terms of value); although the rate fell to 50.3 per cent between 1994 and 1996, the speed of growth remained remarkable.

Of the total amount invested up to the end of 1996, 44 per cent went to South-east Asia, 32 per cent to North America, and 15 per cent to Europe. In terms of projects undertaken, 72 per cent were located in South-east Asia, 14 per cent in North America, and 6 per cent in Europe. By sector, manufacturing accounted for the lion's share of FDI (56 per cent of value and 67 per cent of volume), followed by trade (21 per cent of value and 15 per cent of volume). Together these two sectors accounted for 77 per cent of the total investment made and 82 per cent of the projects undertaken. Within the manufacturing sector, textiles took the largest share of FDI projects (22 per cent), followed by fabricated metals (including electronics, automobiles, and shipbuilding) with a 18 per cent share. However, the fabricated metals sector accounted for the largest proportion of funds invested, with a share of 41 per cent (Bank of Korea 1997: ii–iii, v).

This chapter presents an analysis of the phases and patterns of Korean FDI between 1968 and 1996, identifying the forces driving Korean

Table 5.1 Korean Foreign Direct Investment, 1968–1996

	Net Invested US$mm	*YOY Growth Rate %*	*No. of Cases*
1968–1977	66.9		151
1978	31.4		70
1979	16.4	(47.8)	42
1980	12.3	(25.1)	16
1981	21.9	78.5	36
1982	97.6	345.0	32
1983	102.5	5.0	44
1984	48.2	(53.0)	33
1985	63.8	32.3	9
1986	158.3	148.3	31
1987	320.1	102.2	59
1988	156.2	(51.2)	140
1989	392.4	151.2	246
1990	812.7	107.1	317
1991	1,026.9	26.4	420
1992	1,097.7	6.9	463
1993	1,016.5	(7.4)	618
1994	2,028.8	99.6	1,407
1995	2,756.3	35.9	1,200
1996	3,530.1	28.1	1,319
TOTAL	13,757.0		6,653

YOY Year-on-year

Source: Bank of Korea 1997: 3.

investment, both in terms of 'push' factors in Korea and 'pull' factors in the host markets. The role of the government in promoting FDI is examined, and trends in the location, size, and ownership of overseas projects are highlighted.

A CHRONOLOGICAL OVERVIEW

The history of Korean FDI can be divided into five phases: from the end of the Korean War to the first outward investment from Korea (1953–1967), from 1968 to 1979, from 1980 to 1985, from 1986 to 1996, and from 1997 to the present. This study covers the first four phases of development; the decision to end the survey in December 1996 reflects the author's belief that the economic crisis of 1997 and the reforms implemented as a result of the crisis will have a profound effect on the forms and flows of Korean FDI in the 21st century.

As the *chaebŏl* come under increasing pressure to streamline and rationalize their operations, attention will be paid to the viability of their overseas sales and manufacturing networks. One possible effect of the 'Big Deals' being pushed by the government in 1997–1999 (whereby major conglomerates were required to swap or take over manufacturing affiliates) could be the trimming of overseas operations inherited as part of the deals. The immediate impact of the economic uncertainties in 1997 was a 26 per cent fall in FDI levels in that year to US$2.6 billion (Federation of Korean Banks 1998: 1).

Phase one: 1953–1967

In the immediate post-war period, economic activity in the Republic of Korea focused on import substitution: that is, the importing of capital goods and raw materials to manufacture consumer goods previously sourced from overseas. During this period there was no foreign direct investment to speak of; in 1959, the Korean Traders' Association acquired a piece of commercial real estate in New York City to serve both as the organization's branch office in the United States and as a base for the promotion of Korean exports in North America. (The purchase was financed by foreign exchange earnings from the export of tungsten.) Five years later, a Korean expatriate living in Malaysia set up a timber operation in order to secure a stable supply of lumber for the rapidly-growing Korean plywood industry (Jo 1981: 54–56). However, it was not until the late 1960s that a corporation based in Korea undertook a direct investment project overseas.

Phase two: 1968–1979

The 1960s and 1970s were a period of rapid, export-led economic growth in Korea; the rapid increase in overseas sales encouraged Korean exporters to establish trade offices and processing facilities overseas, while the soaring demand for raw materials prompted overseas investment to secure stable supplies of essential resources. As the import bill for raw materials and intermediate goods swelled, the ROK government instituted an export drive to help offset these costs and encouraged FDI in a restricted number of sectors, including trade and construction.

In 1968 the Korea Development Corporation (KODECO/*Han'guk Nambang Kaebal Chushik Hoesa)* invested US$3.1 million to acquire a 49 per cent share of a joint venture forestry development company in Indonesia (Chŏng 1995: 102). Prior to this investment, Korean individuals and legal entities had not been permitted to make any form of investment overseas (Shin and Oh 1990: 32). However, in 1968, the government enacted a piece of legislation enabling this investment to take place, adding a section governing FDI activity to the Republic of Korea's foreign exchange regulations. For more than a decade after that, the government maintained tight control over outward investment activities, reflecting its desire to control the current account deficit, support the value of the Korean currency, and ensure that payments for imports could be covered by the nation's foreign exchange reserves (Shin Sang-hyup 1998: 46).

Between 1968 and 1979, a total of US$114.7 million was invested in 263 projects on a cumulative, net invested basis. FDI was only permitted in areas beneficial to Korea's industrial development, such as investment aimed at securing stable supplies of raw materials and natural resources needed in the manufacturing and construction sectors (Waitt 1993: 202). Patterns of investment, therefore, showed a fairly even spread among forestry, manufacturing, construction and trade, each of which accounted for between 16 per cent and 21 per cent of the total amount invested (See Table 5.2). In volume terms, investment in trade accounted for the majority share of more than 60 per cent, with 161 projects out of the total of 263. The second ranked sector in terms of volume was fisheries with just 28 projects, or 10.6 per cent of the total.

Table 5.3 shows that, between 1968 and 1979, South-east Asia was the main destination for Korean FDI in terms of value, accounting for 34.2 per cent of cumulative net investment (US$39.2 million), followed by North America (24.1 per cent and US$27.7 million), and Africa (21.0 per cent and US$24.1 million). Together, these three markets accounted for almost 80 per cent of the amount invested overseas by Korean corporations. However, in terms of volume, North America was the favoured host nation with a share of 31.9 per cent (84 out of 263 projects). The second most popular destination was South-east Asia with 72 projects (27.3 per cent), followed by Europe with 46 cases (17.5 per cent). Together, these markets accounted for more than three quarters of Korean investment worldwide.

In the late 1960s and 1970s, Korean foreign direct investment had two principal aims: firstly, to open up markets for Korea's exports (market-

Table 5.2 Korean Foreign Direct Investment: By Sector, 1968–1979

	Net Invested US$mm	*% of Total Net Invested 1968–'79*	*No. of Cases*	*% of Total Cases 1968–'79*
Mining	1.5	1.3	3	1.1
Forestry	22.5	19.6	7	2.7
Fisheries	10.2	8.9	28	10.6
Manufacturing	18.9	16.5	20	7.6
Construction	19.9	17.3	22	8.4
Transportation & Storage	2.0	1.7	9	3.4
Trade	24.2	21.1	161	61.2
Other	15.5	13.6	13	5.0
TOTAL	114.7	100.0	263	100.0

Source: Bank of Korea 1997: 14–15.

Table 5.3 Korean Foreign Direct Investment: By Region, 1968–1979

	Net Invested US$mm	*% of Total Net Invested 1968–'79*	*No. of Cases*	*% of Total Cases 1968–'79*
South-east Asia	39.2	34.2	72	27.3
Middle East	13.5	11.8	22	8.4
Europe	4.5	3.9	46	17.5
North America	27.7	24.1	84	31.9
Latin America	4.2	3.7	17	6.5
Africa	24.1	21.0	17	6.5
Oceania	1.5	1.3	5	1.9
TOTAL	114.7	100.0	263	100.0

Source: Bank of Korea 1997: 26–27.

seeking investment) and, secondly, to secure stable supplies of the raw materials vital to Korea's industrialization programme (resource-seeking investment). In addition, investments were made in construction projects located in the Middle East with the aim of capitalizing on Korea's advantage of abundant, low-cost labour and the experience gained undertaking contracts for the US army during the Vietnam War.

Phase three: 1980–1985

During the first half of the 1980s, Korean FDI remained relatively small in scale, due to a number of factors. The ROK government maintained strict controls over the use of the Republic's limited foreign exchange resources, and Korean corporations lacked the capital, technology and management expertise to invest overseas. Nevertheless, in the six-year period between 1980 and 1985, Korean cumulative net investment amounted to US$346.3 million in 170 projects and the average project size rose to US$2.0 million, up from US$440,000 in the previous decades.

Following the oil shocks of 1973 and 1979, there was a significant increase in investment in the primary sector. This was particularly noticeable in the mining sector, which accounted for half the funds invested overseas (See Table 5.4). Mining and forestry together accounted for 60 per cent of the total investment in this period; the share accounted for by the construction sector declined to less than 5 per cent as the Middle East boom ended, and investment in transportation and storage remained minimal. Manufacturing and trade took smaller shares than before, both in terms of value and volume; together they accounted for 35.0 per cent of the funds invested (down from 37.6 per cent in 1968–1979) and 63.5 per cent of the projects undertaken (down from 68.8 per cent).

Table 5.4 Korean Foreign Direct Investment: By Sector, 1980–1985

	Net Invested US$mm	*% of Total Net Invested 1980–'85*	*No. of Cases*	*% of Total Cases 1980–'85*
Mining	169.7	49.0	10	5.9
Forestry	40.4	11.7	4	2.4
Fisheries	(4.0)		(3)	
Manufacturing	73.8	21.3	51	30.0
Construction	16.1	4.6	23	13.5
Transportation & Storage	0.5	0.1	10	5.9
Trade	47.6	13.7	57	33.5
Other	2.2	0.6	18	10.6
TOTAL	346.3	100.0	170	100.0

Source: Bank of Korea 1997: 14–15.

During this period, South-east Asia lost its position as favoured host nation to North America in terms of value; the region hosted 18.3 per cent of investment compared with 34.2 per cent in the previous period (Table 5.5), while the share accounted for by North America rose from 24.1 per cent to 34.8 per cent. The early 1980s saw a net withdrawal of investment from Africa, leaving Oceania to take third place with US$78.9 million (a 22.8 per cent share of the total). The share of investment accounted for by Europe increased from less than 4 per cent in the 1960s and 1970s to 16.0 per cent by the mid-1980s.

In terms of volume, South-east Asia and North America took similar shares, with 62 and 68 projects respectively. However, there was a marked difference in the size of projects undertaken in the two regions: US$1.0 million in South-east Asia and US$1.8 million in North America. This reflected the fact that investment in South-east Asia was led by small- and medium-sized enterprises engaged in labour-intensive manufacturing activities, whereas investment in North America was spearheaded by the *chaebŏl* in large-scale manufacturing facilities such as electronics factories and automobile assembly plants – early cases of reverse direct investment.

Throughout this period and until the middle of the 1980s, the Korean government regarded FDI as a means of securing stable supplies of resources needed for industrial development at home, boosting sales of Korean manufactured goods, and encouraging co-operation between Korean firms and companies from developing and industrialized nations. The government offered incentives to potential outward investors in target sectors, including protection from investment risk through investment insurance schemes, bilateral investment protection agreements and so forth, financial support in the form of medium- and long-term bank loans, tax incentives, and information-gathering services (Jo 1981: 68).

Table 5.5 Korean Foreign Direct Investment: By Region, 1980–1985

	Net Invested US$mm	*% of Total Net Invested 1980–'85*	*No. of Cases*	*% of Total Cases 1980–'85*
South-east Asia	63.5	18.3	62	36.5
Middle East	26.1	7.5	14	8.2
Europe	55.4	16.0	(3)	
North America	120.6	34.8	68	40.0
Latin America	5.7	1.6	11	6.5
Africa	(3.9)		4	2.4
Oceania	78.9	22.8	14	8.2
TOTAL	346.3	100.0	170	100.0

Source: Bank of Korea 1997: 26–27.

Despite this sector-specific encouragement, FDI was largely restricted to a small number of essential projects in an attempt to prevent the outflow of foreign exchange reserves and a deterioration in the nation's balance of payments situation (Kim Shi-jung *et al* 1992: 29–32). Other factors contributing to the view that FDI should be controlled were official concerns over the possible negative impact of a surge in overseas investment on employment in Korea, often referred to as 'hollowing out' (Kiernan 1994c: 26–27). In 1975, the ROK government set up a system to regulate and administer FDI, covering approval systems, administrative and financial assistance (tax incentives, the conclusion of double taxation agreements and so forth), and follow-up procedures. FDI projects were categorized as 'prohibited', 'suppressed' or 'encouraged'; prohibited projects were those that would have an adverse impact on the economy or which took the form of 'emigration funds' (Shin Sang-hyup 1998: 48–49).

Between 1980 and 1985, changes in the global environment prompted a shift in the patterns of Korean foreign direct investment. As the government's plans to create a strong foundation in the heavy and chemical industries went into effect, Korea's industrial structure became increasingly energy-intensive. The oil shocks and the likelihood of increasing resource nationalism made resource-seeking FDI vital for the nation's continued growth. The economic downturn in Korea in the late 1970s and the prospect of higher wages at home encouraged small- and medium-sized enterprises in labour-intensive manufacturing sectors to transfer some of their production to South-east Asia. The early 1980s saw the beginning of large-scale investment in North America in response to the rising trend of protectionism and the erection of trade barriers. This defensive market-protecting investment was also seen in Europe, resulting in a dramatic increase in the share of global FDI accounted for by investment in Europe.

Phase four: 1986–1996

As has been noted above, the late 1980s and early 1990s saw a spectacular increase in Korean outward investment in terms of volume and value. As late as 1986, there were fewer than 60 new FDI projects per year and, between 1968 and 1985, the total amount invested annually was less than US$100 million. However, the 1990s saw the recording of annual investments in excess of US$1 billion, with more than US$3.5 billion invested in 1996 alone.

In the latter half of the 1980s, the ROK government faced the challenge of managing a rapidly swelling current account surplus and fending off demands from major trading partners for the opening up of Korea's markets to their goods and services. This prompted the government to adopt a more relaxed policy towards FDI; indeed, Waitt observed that criticism of FDI changed into 'almost unqualified support' for outward investment. FDI was now seen as a means of promoting economic development and growth, gaining access to advanced technology, and easing the pressure on Korea's commercial banks to provide funding for investment (Waitt 1993: 202).

The change of policy direction in the late 1980s and early 1990s, from restricting FDI activity to using it as a means of easing trade friction with major trading partners and promoting the globalization of Korean industry, has been reflected in a series of measures implemented over the past decade and aiming to promote FDI by Korean firms (Choi 1996: 223–225). In 1987, the government announced that new investments valued at less than US$1 million and the expansion of existing investments made on the basis of retained earnings could be carried out on a notification basis. This allowed investors to notify the authorities that the investment had taken place, rather than seeking approval before the event.

The following year, the ceiling for notification was raised to US$2 million, making it easier for firms to undertake small-scale investments overseas. The ceiling on projects which had to be referred to the Overseas Investment Screening Council (OISC) for approval was raised in two stages from US$1 million to US$5 million, and the documentation needed to obtain investment approval (where required) and for post-investment administration was simplified. Between 1988 and 1989, the restrictions on the purchase of real estate were eased to allow the acquisition of business-related property as well as land for training facilities and research institutes (Pak Sang-ho 1996: 81–82). The limit on overseas investment by individuals, which had been gradually raised in the late 1980s, was abolished in 1989 (Kim Shi-jung *et al* 1992: 29–32).

In 1990, as the current account went back into deficit (after four years in the black) and there were reports that some overseas investment projects were operating at a loss, the government implemented a partial revision of the FDI approval system, which included the strengthening of the supervision of large-scale projects. However, two years later, the process of deregulation resumed with the ceilings on projects that could be undertaken on a notification basis and on projects to be referred to the OISC being raised to US$5 million and US$10 million respectively. In 1992, the government also abolished the requirement that companies planning FDI projects should use their own (i.e. internally-generated) funding when investing overseas.

The following year, the government reduced the number of sectors in which FDI was restricted to 17, and allowed the purchase of real estate overseas for the purpose of building housing for workers. Approval for activities such as the leasing of overseas real estate and the building and operating of golf ranges, which had previously been restricted to construction firms with permits for overseas construction or General Trading Companies, were opened to all companies and individuals. The Industrial Bank of Korea and the Small and Medium Industry Bank were given the authority to approve FDI projects, expanding the number of financial institutions from whom potential investors could seek approval. As a result of these changes in the system, FDI projects valued at less than US$10 million (which made up more than 95 per cent of Korean overseas investment) were effectively completely liberalized. Projects valued at between US$10 million and US$50 million could simply be reported to a designated institution, while projects valued at more than US$50 million could proceed subject to official approval (Ch'oi 1997: 66–67).

Between 1986 and 1996, Korean companies invested US$13.3 billion in more than 6,200 projects worldwide. The early 1990s saw an explosion in Korean FDI; in the previous decade, the annual amount invested overseas by Korean corporations had fluctuated between US$12 million and US$392 million (See Table 5.1). In 1991, Korean outward FDI passed the US$1 billion mark and remained at that level for the next two years, before breaking through the US$2 billion level in 1994 and the US$3 billion mark in 1996. In one decade, annual net investment rose from US$320 million to US$3.5 billion, and the cumulative net amount rose from US$619.3 million to US$13.8 billion (Bank of Korea 1997: 15).

The majority of both the value (US$8.3 billion or 60.1 per cent) and volume (3,926 projects or 59.0 per cent) of Korean FDI was invested between 1994 and 1996. In 1996 alone, Korean FDI totalled US$3.5 billion in 1,319 projects, accounting for 25.4 per cent and 19.8 per cent of the total cumulative value and volume respectively. Despite this rapid increase in investment activity, Korean FDI remained at a low level in terms of its share of gross national product. In 1995, Korean FDI was equivalent to just 2.3 per cent of GNP compared with between 7.0 per cent and 10.0 per cent in the United States and Japan, and 24.0 per cent in the United Kingdom (*Business Korea* March 1996: 17–21).

The most recent period of expansion of Korean FDI began in 1994, the year in which the government implemented a complete overhaul of the investment approval system. The new focus on the promotion of globalization, together with concerns over increasing regionalism and intensifying global competition encouraged the government to regard FDI as a key part of a strategy of growth and survival (Lee Chae-min 1997: 4–5).

The old system of designating encouraged, restricted and general sectors was simplified to a list of restricted sectors, including industries in which FDI could lead to weakened competitiveness for Korean industry back at home, where there was a possibility of damage from reverse imports, or where overseas investment could lead to excessive competition among Korean firms. Projects up to the value of US$10 million could proceed on the basis of notification to an even larger number of financial institutions (including all foreign exchange banks) and, in future, only projects valued at more than US$10 million would be subject to deliberation and approval by the competent authorities (Han To-suk 1996: 81–82 and Pak Sang-ho 1995: 182–195).

The rapid inflow of foreign funds witnessed in 1995 as market liberalization programmes were implemented prompted the ROK government to take further action to promote FDI. This, it was hoped, would serve as a counterbalance to the inflow of funds and reduce the possibility of an acceleration in inflationary pressures. The authorities issued a 'Directive for the Liberalization and Stabilization of Foreign Direct Investment' as part of the preparations to meet the capital requirements for becoming a member of the OECD. The ceiling for notification was raised substantially to US$50 million (with projects in excess of this amount to be referred to the OISC as before), and the number of restricted sectors was further reduced from 14 to three.

However, the requirement for internal funding was re-introduced, stipulating that investors should find 10 per cent of the investment capital needed for an FDI project valued up to US$100 million and 20 per cent of amounts for larger projects from their own capital. This measure aimed to deter 'reckless' investment, ensure the sound operation of overseas ventures, and protect Korea from an increase in foreign debt should an overseas investment fail (*Business Korea* March 1996: 17–21).

1996 saw the liberalization of the last three restricted sectors, a further simplification of approval and supervision procedures, and the granting of automatic approval within 15 days for investments in countries with an A credit rating (Pak Sang-ho 1995: 182–195 and Han To-suk 1996: 81–82). The following year, the government once again abolished the self-financing requirement, allowing firms to finance FDI projects entirely with external funding and without any limit on the amount invested (subject to the consent of their major creditor bank).

The government allowed a further easing of the regulations to permit any investment valued at less than US$50 million to proceed upon notification to a major creditor bank, rather than to the Bank of Korea as had been the case in the past. The OISC would continue to screen certain cases where investment exceeded US$50 million, including cases where the aggregate amount invested overseas exceeded 50 per cent of the mother company's net worth or where the investment was to be made by a 'capital challenged' firm (Han To-suk 1996: 81– 82).

In the latter half of the 1990s, the ROK government gave priority in approval, notification and support to investments in six categories: investment aimed at developing or acquiring advanced technology, relocating production facilities from industries with a weak competitive advantage, securing resources and raw materials, expanding export markets, and making a direct contribution to the promotion of economic co-operation with foreign countries – or any investment designated 'necessary' by the OISC (Pak Sang-ho 1995: 195).

As Table 5.6 shows, recent years have seen a rapid decline in the share of outward FDI accounted for by the primary sector, which fell to 6.3 per cent in terms of value and 2.9 per cent in terms of volume (compared with 60.7 per cent and 8.3 per cent respectively between 1980 and 1985). There has been a corresponding increase in the share taken by manufacturing and trade, which together accounted for 78.1 per cent of the amount invested and 82.8 per cent of the projects undertaken. The manufacturing sector took the largest share of investment (US$7.6 billion or 57.4 per cent) and over 70 per cent of the volume (4,388 projects). After manufacturing and trade, the next largest shares in value terms were taken by mining (5.5 per cent) and construction (3.7 per cent). Once again, only a small percentage of investment (0.7 per cent) was made in the transportation and storage sector.

Between 1986 and 1996, South-East Asia regained its position as most favoured host country for Korean investment, accounting for 44.8 per cent

Table 5.6 Korean Foreign Direct Investment: By Sector, 1986–1996

	Net Invested US$mm	*% of Total Net Invested 1986–'96*	*No. of Cases*	*% of Total Cases 1986–'96*
Mining	792.2	5.5	52	0.8
Forestry	21.0	0.2	10	0.2
Fisheries	83.8	0.6	118	1.9
Manufacturing	7,630.5	57.4	4,388	70.5
Construction	491.7	3.7	163	2.6
Transportation & Storage	88.5	0.7	69	1.1
Trade	2,747.2	20.7	764	12.3
Other[a]	1,504.0	11.2	656	10.6
TOTAL	13,295.9	100.0	6,220	100.0

[a] No breakdown is given for 'other' investments.

Source: Bank of Korea 1997: 14–15.

of the amount invested (US$6.0 billion) and 75.1 per cent of the projects undertaken (4,672 cases). The average amount per project in the region was US$1.3 million, as small- and medium-sized enterprises continued to lead investment in small-scale manufacturing facilities. North America took second position in terms of both value (US$4.2 billion, 31.4 per cent) and volume (768 projects, 12.3 per cent); the average project size, at US$5.4 million, was higher than in South-east Asia as investment focused on large-scale, capital-intensive facilities.

Investment in Europe rose sharply ahead of the formation of a single market in 1992, with Europe rising to third place in terms of value (US$2.0 billion, 15.4 per cent) and volume (341 projects, 5.5 per cent), with an average value per investment of US$5.6 million (See Table 5.7).

Whereas, in the early stages of Korean FDI activity, Korea's small- and medium-sized enterprises had taken the lead, from 1993 larger companies began to assume a dominant position, especially in the semiconductors, electronics and automobiles sectors. In 1993, SMEs accounted for 87.2 per cent of investment volume and 72.8 per cent of the value; by 1995 these shares had dropped to 68.0 per cent and 15.0 per cent respectively. In the 1990s, large-scale investment in industrialized countries was being undertaken principally by the *chaebŏl* in leading edge industries to secure market share, acquire technology, circumvent protectionist measures, and collect market information. In contrast, small-scale investment in developing countries was being made by SMEs in sunset industries, mainly with the aim of exporting back to Korea (Cho 1996: 17); only 20 per cent of the output in developing countries was sold locally (Pak Sang-ho 1996: 46–47).

As Korea moved into the 'Boom Years' of the late 1980s, changes were already taking place that would have a profound impact on the Korean economy, its leading industries, and FDI activity by major Korean

Table 5.7 Korean Foreign Direct Investment: By Region, 1986–1996

	Net Invested US$mm	*% of Total Net Invested 1986–'96*	*No. of Projects*	*% of Total Projects 1986–'96*
South-east Asia	5,952.7	44.8	4,672	75.1
Middle East	88.7	0.7	3	0.1
Europe	2,044.3	15.4	341	5.5
North America	4,178.7	31.4	768	12.3
Latin America	546.1	4.1	240	3.9
Africa	253.7	1.9	36	0.6
Oceania	231.7	1.7	160	2.5
TOTAL	13,295.7	100.0	6,220	100.0

Source: Bank of Korea 1997: 26–27.

corporations. The late 1980s and early 1990s saw significant changes in the flow and forms of Korean FDI, as the focus shifted away from the search for new markets and sources of raw materials and energy (market- and resource-seeking investment) towards the defence of existing markets, procurement of sites for low-cost production, and the acquisition of technology and expertise (efficiency- and strategic capability-seeking investment).

In the 1970s and early 1980s, Korean companies had focused on exports and sales in the domestic market where, in many cases, they enjoyed protection from foreign competition. As has been discussed above, in export markets they enjoyed a strong price advantage deriving from favourable production conditions at home. However, the changes in the domestic and global business environment in the latter half of the 1980s put FDI into an entirely different perspective.

Indeed, the move overseas must have appeared to many companies to be an inevitable step in their efforts to counteract these changes; at home, they faced an erosion of their competitive advantage due to the appreciation of the wŏn, rising wages and land prices, falling productivity, and labour unrest. Internationally, they had to overcome trade friction and import restrictions, and formulate a response to the emergence of regional economic blocs such as the North America Free Trade Association (NAFTA) and the European Union (EU).

For many companies, survival meant relocating production facilities overseas in developing countries to achieve lower production costs, and in industrialized countries to combat protectionism and acquire advanced technology and management expertise. Therefore, FDI came to be seen as an integral part of the industrial restructuring process and an important element in their globalization strategy. This was, clearly, a key motivation behind the FDI boom of recent years. Another significant factor was the establishment of diplomatic and economic relations with former communist countries, opening up huge potential markets in Asia, and central and eastern Europe.

Scale of investment and ownership ratios

At the end of 1979, out of a total cumulative investment of US$114.7 million in 263 projects, 238 projects (90.5 per cent of the total) were valued at less than US$1 million. Five projects (1.9 per cent) were valued at more than US$5 million, and only one (0.4 per cent) was valued at more than US$10 million. Between 1968 and 1979, the average amount invested per project was US$436,200, and by the end of 1985, the majority of projects were still valued at less than US$1 million (368 cases out of 433, or 85.0 per cent). A slightly higher proportion of projects (11 or 2.5 per cent) were valued at more than US$5 million, and six projects (1.4 per cent) were valued at more than US$10 million. By the mid-1980s, the average investment per project had risen slightly to US$1.1 million.

The trend seen in the early 1980s continued into the latter half of the 1990s, with a slight decline in the share of small projects. Out of a total of 6,653 projects outstanding at the end of 1996, just over three quarters (5,107 projects or 76.7 per cent) were valued at less than US$1 million. Larger investments now accounted for a higher share of the total, with 200 (3.0 per cent) valued at more than US$5 million and 220 (3.3 per cent) valued at more than US$10 million. By the end of 1996, the average amount invested per project that year had risen to US$2.1 billion. The statistics show an increase of almost five times in the average size per project and a small, but rising proportion of large-scale projects, reflecting the increasing participation in large-scale FDI projects by major firms (Bank of Korea 1997: 73).

FDI statistics reveal that, although Korean investors have historically preferred wholly-owned enterprises to joint ventures, recent years have seen an increasing liking for co-operative projects. At the end of 1979, around 20 per cent of Korean investment projects (54 projects out of 263) had an equity share of 50 per cent or less. By 1985, the share had risen to 26.3 per cent (114 projects out of 433) and a decade later more than 28 per cent of Korean investment projects were operating on the basis of a minority share holding (1,870 projects out of 6,653). However, majority share holdings still accounted for more than 70 per cent of Korean investments, reflecting a continuing preference for higher levels of management control.

Nevertheless, the figures show an increasing willingness for some foreign participation in projects; the share of the total accounted for by wholly-owned enterprises has fallen from 72.7 per cent at the end of 1979 to 63.4 per cent in 1985 and to 51.0 per cent in 1996 (Bank of Korea 1997: 75). In the past, Korean investors have tended to demonstrate a preference for 100 per cent-owned enterprises or majority share joint ventures, even when they lacked overseas management experience. In previous years, they had felt the need for strong control in order to reduce the management costs that arose due to cultural differences. More recently, however, they have been entering into more joint ventures in order to find a partner to help them break into the local market, obtain information, or establish an advance market for regional sales (Wang 1997: 63–72).

Worthy of comment is the recent increase in mergers and acquisitions (M&A) as a means of FDI; once again the motivation for investment being to avoid trade barriers, acquire technology, obtain famous brands and thus raise consumer awareness, or expand market power. The trend has been particularly noticeable in high-tech areas such as computers and semi-conductors, with acquisitions made by the larger conglomerates. Korean analysts note that, given the short history of M&A activity, it is hard to say whether such activity has been a success (Lee Chae-min 1997: 15).

KOREAN MANUFACTURING INVESTMENT

During the past three decades, by far the greatest focus of Korean investors has been the manufacturing sector, which has accounted for 56.1 per cent (US$7.7 billion) of the total amount invested and 67.0 per cent of projects (4,459) undertaken overseas. As can be seen in Table 5.8, trade and manufacturing investment have together accounted for more than three quarters of the value and in excess of 80 per cent of the volume of Korean FDI since 1968.

Although Korean foreign direct investment in the manufacturing sector began in 1973, during the 1970s the focus of Korean FDI was more on resource development and trade rather than manufacturing. Between 1968 and 1979, investment in the manufacturing sector amounted to just US$19 million on a cumulative net invested basis. The manufacturing sector accounted for less than 1 per cent of all funds invested and projects undertaken up to 1979 (See Table 5.9). At the end of 1985, manufacturing investment continued to account for only a small share of Korean investment value (1.0 per cent) and volume (1.1 per cent). However, investment in the manufacturing sector began to accelerate from 1986,

Table 5.8 Korean Foreign Direct Investment: By Sector, 1968–1996

	Net Invested US$mm	*% of Total Net Invested 1968–'96*	*No. of Cases*	*% of Total Cases 1968–'96*
Mining	900.3	6.5	65	1.0
Forestry	83.9	0.6	21	0.3
Fisheries	90.0	0.7	143	2.1
Manufacturing	7,723.3	56.1	4,459	67.0
Construction	527.7	3.8	208	3.1
Transportation & Storage	90.9	0.7	88	1.3
Trade	2,819.0	20.5	982	14.8
Other	1,521.8	11.1	687	10.4
TOTAL	13,757.0	100.0	6,653	100.0

Source: Bank of Korea 1997: 14–15.

Table 5.9 Korean Manufacturing Investment, 1968–1996

	Net Invested US$mm	*% of Total Net Invested*	*Cases*	*% of Total Cases*
1968–1979	19.0	0.2	20	0.4
1980–1985	73.8	1.0	51	1.1
1986–1996	7,630.5	98.8	4,388	98.5
TOTAL	7,723.3	100.0	4,459	100.0

Source: Bank of Korea 1997: 14.

Table 5.10 Korean Manufacturing Investment: By Sector and Region, December 1996

	Mining US$mm (Cases)	*Lumber US$mm (Cases)*	*Fishing US$mm (Cases)*	*Manuf. US$mm (Cases)*	*Constr. US$mm (Cases)*	*Tran./Sto. US$mm (Cases)*	*Trade US$mm (Cases)*	*Other US$mm (Cases)*	*TOTAL US$mm (Cases)*
SE Asia	344	31	18	4,732	327	62	392	365	6,371
	(21)	(8)	(59)	(3,803)	(125)	(56)	(397)	(361)	(4,835)
North America	119	Neg.	21	1,598	126	15	1,717	732	4,327
	(10)	(1)	(12)	(286)	(41)	(19)	(340)	(211)	(920)
Latin/Cent. America	77	6	45	329	3	12	46	38	556
	(4)	(2)	(53)	(154)	(4)	(8)	(32)	(11)	(268)
Europe	89	1	3	916	12	3	618	247	1,888
	(4)	(2)	(7)	(126)	(5)	(5)	(163)	(45)	(355)
Middle East	148	0	0.2	25	13	0	3	73	261
	(5)	(0)	(2)	(17)	(16)	(0)	(6)	(6)	(52)
Africa	0	0	1	82	34	0	11	14	141
	(0)	(0)	(3)	(19)	(3)	(0)	(12)	(7)	(44)
Oceania	123	46	2	42	13	0	33	53	312
	(21)	(8)	(7)	(54)	(14)	(0)	(27)	(48)	(179)
TOTAL	900	84	90	7,723	528	91	2,819	1,522	13,757
	(65)	(21)	(143)	(4,459)	(208)	(88)	(982)	(687)	(6,653)

Source: Ha and Hong 1998: 22.

increasing rapidly after 1994. The statistics show that almost all Korean manufacturing investment was made between 1986 and 1996, with nearly two thirds of projects being initiated between 1994 and 1996 (US$5.0 billion out of US$7.7 billion) (Bank of Korea 1997: 14).

In the late 1980s and early 1990s, North America and South-east Asia continued to account for the largest share of South Korea's total FDI; these two markets were also host to the greatest proportion of manufacturing investment (See Table 5.10). However, as late as 1987, Korean FDI was centred on North America, and it was only in the latter half of the 1980s

Table 5.11 Korean Manufacturing Investment: By Top Five Host Countries, December 1996

	1st US$mm	*2nd US$mm*	*3rd US$mm*	*4th US$mm*	*5th US$mm*
Food & Beverages	PRC	INDO	USA	UK	HK
	133	114	25	15	10
Textiles & Clothing	PRC	INDO	VIET	UZB	USA
	340	73	68	63	51
Footwear & Leather	PRC	INDO	VIET	SUD	THAI
	174	40	36	16	14
Wood & Furniture	INDO	PRC	USA	CIS	MAL
	40	39	36	16	11
Paper & Printing	THAI	PRC	USA	CAN	INDO
	35	33	27	15	4
Petroleum & Chemicals	PRC	INDO	USA	HK	PHIL
	233	114	50	31	31
Non-ferrous Metals	PRC	MAL	USA	INDO	PHIL
	186	56	54	36	14
Primary Metals	USA	CAN	PRC	VIET	CIS
	416	113	63	19	15
Fabricated Metals	USA	PRC	INDO	IND	VIET
	663	623	183	179	160
Machinery & Equipment	PRC	USA	UK	INDO	BEL
	169	43	22	17	13
Other	PRC	USA	INDO	MAL	HK
	250	75	46	43	32
TOTAL	PRC	USA	INDO	VIET	MAL
	2,243	1,448	677	335	285

BEL	Belgium	PHIL	Philippines
CAN	Canada	PRC	People's Republic of China
CIS	Commonwealth of Independent States	SUD	Sudan
		THAI	Thailand
HK	Hong Kong	UK	United Kingdom
IND	India	USA	United States of America
INDO	Indonesia	UZB	Uzbekistan
MAL	Malaysia	VIET	Vietnam

Source: Ha and Hong 1998: 25.

that SMEs in the light industrial sectors (textiles, clothing, footwear, toys and so forth) began to invest heavily in South-east Asia in an effort to restore their export price competitiveness.

A noteworthy development has been the diversification of FDI into new markets made accessible by the collapse of communism and the opening up of the eastern bloc. The 'Northern Policy' adopted by the ROK government in the late 1980s, which culminated in the establishment of diplomatic relations with the People's Republic of China and the Soviet Union in the early 1990s, opened up attractive new markets for Korean manufacturing investment. Similarly, the collapse of communism in central and eastern Europe opened the door for full diplomatic and economic links between Korea and the former eastern bloc states.

Table 5.11 shows the impact of the opening up of these new markets; at the end of 1996, the People's Republic of China accounted for the largest share of Korean FDI in six out of ten manufacturing sectors (excluding 'other investments'), namely food and beverages, textiles and clothing, footwear and leather, petroleum and chemicals, non-ferrous metals, and machinery and equipment. The United States led the field in just two sectors: primary metals and fabricated metals. The remaining sectors (wood and furniture, and paper and printing) were dominated by Indonesia and Thailand. Vietnam ranked among the top five host countries for four sectors (textiles and clothing, footwear and leather, primary and fabricated metals) to come fourth overall.

Over the past three decades, manufacturing investment by Korean companies has been led by fabricated metals, with a 41.0 per cent share in value terms (See Table 5.12). There is a significant gap of more than US$2

Table 5.12 Korean Manufacturing Investment: By Sector, December 1996

	Value US$mm	*% of Total Value*	*Cases*	*% of Total Cases*
Fabricated Metals	3,169.2	41.0	817	18.3
Textiles & Clothing	949.2	12.3	965	21.6
Primary Metals	703.1	9.1	168	3.8
Petroleum & Chemicals	657.9	8.5	408	9.2
Non-ferrous Metals	397.9	5.2	175	3.9
Foods & Beverages	350.4	4.5	301	6.8
Footwear & Leather	321.9	4.2	378	8.5
Machinery & Equipment	297.4	3.9	199	4.5
Wood & Furniture	160.3	2.1	186	4.2
Paper & Printing	126.9	1.6	84	1.9
Other	89.1	7.6	778	17.4
TOTAL	7,723.3	100.0	4,459	100.0

Source: Bank of Korea 1997: 83.

Table 5.13 Korean Manufacturing Investment: By Sector and Region, December 1996

	SE Asia US$mm (%)	*Middle East US$mm (%)*	*N. America US$mm (%)*	*L. America US$mm (%)*	*Europe US$mm (%)*	*Africa US$mm (%s)*	*Oceania US$mm (%)*	*TOTAL US$mm*
Food & Beverages	279.3	0.6	26.5	8.6	24.6	0.0	10.9	350.4
	(79.7)	(0.2)	(7.6)	(2.5)	(7.0)	(0.0)	(3.1)	
Textiles & Clothing	613.6	0.4	51.3	119.2	97.5	52.8	14.3	949.2
	(64.6)	(0.04)	(5.4)	(12.6)	(10.3)	(5.6)	(1.5)	
Footwear & Leather	282.6	0.0	7.6	3.0	7.8	18.9	2.0	321.9
	(87.8)	(0.0)	(2.4)	(0.9)	(2.4)	(5.9)	(0.6)	
Wood & Furniture	92.3	0.0	45.7	4.4	216.2	0.0	1.7	160.3
	(57.6)	(0.0)	(28.5)	(2.7)	(10.1)	(0.0)	(1.1)	
Paper & Printing	83.0	0.0	41.7	0.3	1.9	0.0	0.0	126.9
	(65.4)	(0.0)	(32.9)	(0.2)	(1.5)	(0.0)	(0.0)	
Petroleum & Chemicals	488.6	13.1	53.1	26.7	68.6	7.7	0.1	657.9
	(74.3)	(2.0)	(8.1)	(4.1)	(10.4)	(1.2)	(0.01)	
Non-ferrous Metals	325.2	0.0	54.0	0.8	9.7	0.3	7.9	397.9
	(81.7)	(0.0)	(13.6)	(0.2)	(2.4)	(0.1)	(2.0)	
Primary Metals	113.7	3.5	528.9	27.8	28.8	0.0	0.4	703.1
	(16.2)	(0.5)	(75.2)	(4.0)	(4.1)	(0.0)	(0.1)	
Fabricated Metals	1,560.2	2.5	669.4	123.4	806.1	3.6	4.0	3,169.2
	(49.2)	(0.1)	(21.1)	(3.9)	(25.4)	(0.1)	(0.1)	
Machinery & Equipment	213.12	0.0	42.9	0.0	41.1	0.2	0.0	297.4
	(71.7)	(0.0)	(14.4)	(0.0)	(13.8)	(0.1)	(0.0)	
Other	467.6	1.6	76.4	15.1	26.5	1.0	1.0	589.0
	(79.4)	(0.3)	(13.0)	(2.6)	(4.5)	(0.2)	(0.2)	
TOTAL	4,519.1	21.7	1,597.6	329.2	1,128.8	84.6	7,723.3	4,519.1

Source: Bank of Korea 1997: 82–83.

billion between the fabricated metals sector and the next ranked group, textiles and clothing, which took a 12.3 per cent share. Primary metals and petroleum and chemicals ranked third (9.1 per cent) and fourth (8.5 per cent) respectively.

As can be seen in Table 5.13, South-east Asia accounted for 87.8 per cent of South Korea's total investment in the footwear and leather sector, 81.7 per cent in non-ferrous metals, and 79.7 per cent in food and beverages. The region's shares for other sectors were also high: 74.3 per cent for petroleum and chemicals, 71.7 per cent for machinery and equipment, 65.4 per cent for paper and printing, 64.6 per cent for textiles and clothing, and 57.6 per cent for wood and furniture. South-east Asia took a share of less than 50 per cent in just two sectors, namely fabricated metals (49.2 per cent) and primary metals (16.2 per cent). However, with a cumulative investment value of US$1.6 billion, investment in the fabricated metals sector accounted for 34.5 per cent of Korean investment in the region with a value of one and a half times the next sector, textiles and clothing (US$613.6 million).

North America accounted for the lion's share of just one sector, namely primary metals (75.2 per cent); other significant shares were in paper and printing (32.9 per cent) and wood and furniture (28.5 per cent). Once again, although only 21.1 per cent of global investment in fabricated metals went to North America, in value terms the sector accounted for more than 40 per cent of Korean FDI in the region. Fabricated metals and primary metals together took a 75 per cent share of all Korean investment in North America. There was little investment (6 per cent or less) in labour-intensive sectors such as textiles and clothing, footwear and leather, which was to be expected in a high labour cost market. Europe did not take a leading share in any sector; the highest share was for fabricated metals (25.4 per cent), followed by machinery and equipment (13.8 per cent). The pattern seen in other countries was repeated, with the fabricated metals sector accounting for more than 70 per cent of all Korean investment in Europe.

An analysis of Korean FDI at the end of the 1990s by Ha and Hong revealed changes in the distribution of investment by large firms and SMEs (See Table 5.14). As of the end of 1996, SMEs accounted for 66.8 per cent of the volume of investment (4,507 cases) but only 21.6 per cent of the amount invested (US$3.0 billion). Also worthy of note is the increase in FDI by individuals, which increased from just one instance in 1987 to almost 350 projects a decade later (Ha and Hong 1998: 26).

INVESTMENT IN INDUSTRIALIZED COUNTRIES AND DEVELOPING COUNTRIES

In a report published by the Korea Development Institute in 1998, Kim Sŭng-jin analyzed the distribution of Korean FDI among industrialized or advanced country (AC) markets and developing country (DC) markets.[2]

Table 5.14 Korean Foreign Direct Investment Projects: By Large Enterprises and SMEs, December 1996

	Large Firms	*SMEs*	*Individuals*	*TOTAL*
1985	27	11	0	38
1986	36	14	0	50
1987	58	32	1	91
1988	110	60	2	172
1989	122	139	8	269
1990	144	185	10	339
1991	152	282	9	443
1992	128	366	3	497
1993	134	543	6	683
1994	252	1,146	77	1,475
1995	261	918	112	1,291
1996	232	811	349	1,392
TOTAL	1,656	4,507	577	6,740
% of TOTAL	24.6	66.8	8.6	100.0

Source: Ha and Hong 1998: 26.

Out of a cumulative total of US$13.8 billion, 54.0 per cent (US$7.4 billion) was invested in DC markets, 45.1 per cent (US$6.2 billion) in AC markets, and 0.9 per cent (US$128 million) in the Middle East (See Table 5.15).

In the manufacturing sector, however, the distribution was far less evenly balanced, with 71.5 per cent (US$5.5 billion out of the total of US$7.7 billion) invested in DC markets, 28.2 per cent (US$2.2 billion) in AC markets, and 0.3 per cent (US$22 million) in the Middle East. An analysis by sector revealed that 39.9 per cent of Korean investment in AC markets was in the trade sector, with 35.1 per cent invested in manufacturing. In contrast, 74.4 per cent of funds invested in DC markets went into manufacturing, with less than 10 per cent going to each of the other sectors. These figures reflect the nature of Korean FDI in each area; investment in trade assumed a greater importance in markets such as North America and western Europe (market-seeking investment), while investment in DC markets sought lower production costs and was, therefore, focused on manufacturing activity.

Table 5.16 shows that more than 90 per cent of Korean investment in two sectors (textiles and clothing, footwear and leather), over 80 per cent of investment in a further three sectors (petroleum and chemicals, non-ferrous metals, and food and beverages), more than 70 per cent in wood and furniture, and in excess of 66 per cent of investment in fabricated metals went to DC markets. More than 77 per cent of investment in the primary metals sector went to AC markets, but this was the only sector in which FDI was focused on industrialized countries. The second ranking investment sector in AC markets was fabricated metals, with a 33 per cent share of investment.

Table 5.15 Korean Foreign Direct Investment: By Sector and AC/DC Markets, December 1996

	AC *US$mm (%)*	*DC* *US$mm (%)*	*Middle East* *US$mm (%)*	*TOTAL* *US$mm*
Primary	319.0 (5.1)	665.2 (9.0)	90.0 (70.2)	1,074.2
Manufacturing	2,177.5 (35.1)	5,524.1 (74.4)	21.7 (16.9)	7,723.3
Construction	141.1 (2.3)	373.4 (5.0)	13.2 (10.3)	527.7
Transportation & Storage	25.3 (0.4)	65.6 (0.9)	0.0 (0.0)	90.9
Trade	2,477.9 (39.9)	339.3 (4.6)	1.8 (1.4)	2,819.0
Other	1,062.9 (17.1)	547.4 (6.2)	1.5 (1.2)	1,521.8
TOTAL	6,203.8 (45.1)	7,425.0 (54.0)	128.2 (0.9)	13,757.0

Source: Kim Sŭng-jin 1998: 27.

Table 5.16 Korean Manufacturing Investment: By Sector and AC/DC Markets, December 1996

	AC *US$mm (%)*	*DC* *US$mm (%)*	*Middle East* *US$mm (%)*	*TOTAL* *US$mm*
Food & Beverages	52.5 (15.0)	297.3 (84.8)	0.6 (0.2)	350.4 (100.0)
Textiles & Clothing	60.0 (6.3)	888.8 (93.7)	0.4 (Neg.)	949.2 (100.0)
Footwear & Leather	17.0 (5.3)	304.9 (94.7)	0.0 (0.0)	321.9 (100.0)
Wood & Furniture	46.2 (28.8)	114.1 (71.2)	0.0 (0.0)	160.3 (100.0)
Paper & Printing	43.6 (34.4)	83.3 (65.6)	0.0 (0.0)	126.9 (100.0)
Petroleum & Chemicals	98.7 (15.0)	546.1 (83.0)	13.1 (2.0)	657.9 (100.0)
Non-ferrous Metals	67.7 (17.0)	330.2 (83.0)	0.0 (0.0)	397.9 (100.0)
Primary Metals	543.1 (77.2)	156.5 (22.3)	3.5 (0.5)	703.1 (100.0)
Fabricated Metals	1,151.2 (33.2)	2,312.9 (66.7)	2.5 (0.1)	3,466.6 (100.0)
Other	97.5 (16.6)	489.9 (83.1)	1.6 (0.3)	589.0 (100.0)
TOTAL	2,177.5 (28.2)	5,524.0 (71.5)	21.7 (0.3)	7,723.3 (100.0)

Source: Kim Sŭng-jin 1998: 27.

Table 5.17 Korean Exports and FDI: Leading Sectors, 1990 & 1994

	1990			
	1st	*2nd*	*3rd*	*4th*
Exports	FM	T&C	F&L	P&C
% of Total	40.6	24.0	9.9	8.6
FDI	FM	PM	T&C	P&C
% of Total	26.8	22.9	13.5	8.6

	1994		
	1st	*2nd*	*3rd*
Exports	FM	T&C	P&C
% of Total	48.5	19.8	13.1
FDI	FM	T&C	P&C
% of Total	38.8	13.0	9.7

Source: Kim Sŭng-jin 1998: 32.

Kim's analysis of FDI activity also revealed a clear relationship between FDI and exports; in 1990 and 1994, the fabricated metals sector accounted for the lion's share of both exports and FDI. The textiles and clothing sector, which took second place in export performance in both years, was the third largest FDI sector in 1990, rising to second largest in 1994. By that year, the petroleum and chemicals sector, which had taken fourth place in both exports and FDI in 1990, had moved up to third place in both areas of performance. The statistics clearly show a relationship between Korea's top three export sectors and FDI activity. The top three export sectors, which accounted for 74.5 per cent of all exports in 1990, also accounted for 40.3 per cent of FDI. In 1994, the top three export sectors accounted for 81.4 per cent of exports and 61.5 per cent of FDI (See Table 5.17).

CONCLUSIONS

An analysis of the various phases of Korean foreign direct investment between 1968 and 1996 reveals a number of important points. Firstly, government policy has played a crucial role in either curbing or promoting outward investment, according to the overall economic strategy in place at the time. In the 1960s and 1970s, although Korean companies were encouraged to invest overseas, they were restricted to investments that secured stable supplies of vital resources, or opened up and expanded new markets for Korean exports. The overriding concerns for the government were the maintenance of appropriate levels of foreign exchange reserves, the reduction of the current account deficit, and the protection of employment at home.

However, as we have already seen, changes in the domestic and international business environment in the late 1980s and 1990s led to a sea change in the perception of FDI by government and corporations alike. Overseas investment came to be seen as a means of easing friction with major trading partners (by creating jobs in the host country), enhancing export price competitiveness (by relocating manufacturing facilities in developing countries), developing non-price competitiveness (by acquiring advanced technology and expertise through collaborative agreements, joint ventures or M&A), and achieving the goal of globalization (by establishing international research, sales, production and distribution networks). The government's decision to actively promote FDI came at a time when Korean corporations were becoming acutely aware of the need for overseas investment.

The late 1980s and early 1990s also saw a change in the location of Korea FDI that reflected the altered view of its role in the economy. The focus of investment switched from resource-seeking investment in developing countries in the 1970s and early 1980s, to market-seeking (or even market-defending) investment in industrialized countries in the late 1980s and early 1990s. The most recent development has been the expansion of efficiency-seeking investment, such as production in eastern Europe for distribution in the EU, and strategic capability-seeking investment in industrialized countries, including the development of design and research capabilities in western Europe and the United States.

Korean FDI has been dominated by investment in manufacturing and trade. As might be expected, labour-intensive production facilities have been located in developing countries, with capital- and technology-intensive manufacturing taking place in industrialized countries. Although the focus of investment in the trading sector has been the more prosperous industrialized nations, the Koreans have also been establishing sales and distribution networks in developing countries (most recently the countries of central and eastern Europe) where medium- and long-term prospects for demand are favourable.

In the early stages of Korean FDI, small- and medium-sized enterprises played a leading role in the advance overseas; the emergence of the *chaebŏl* as major investors is a more recent phenomenon. The increasing involvement of big business in overseas investment is indicated by the gradual rise in the average size of FDI projects and the undertaking of multi-million dollar projects in industrialized countries. Over the past decade, the *chaebŏl* have clearly realized the importance of becoming global players – competing with each other in export markets from the safety of their home territory is no longer an option. In an era of unlimited competition, the Korean conglomerates face tough competition in the domestic market as well as overseas.

It is the author's belief that the shifts in the nature and patterns of Korean FDI and the changes in attitudes towards overseas investment outlined above are due, in large part, to the changes in the domestic and global business environment that have occurred since the late 1980s. Korea's FDI boom has been a response to those changes; since the latter half of the 1980s, the ROK government has played a significant role in promoting FDI as a means of achieving its broader economic goals. For their part, Korean corporations have engaged in FDI as a response both to 'push' factors at home (such as rising production costs and weakening competitiveness) and 'pull' factors overseas. As many studies have already shown, Korean investors have been attracted by the favourable conditions in host countries, with industrialized countries seeking inward investment to create new jobs in their own manufacturing industries, and developing countries implementing policies to attract foreign capital in support of their own economic development programmes.

This chapter has examined the principal factors that have driven Korean global FDI over the past three decades. In the following chapters, Korean investment in Europe will be analyzed within the context of Korea's evolving relationship with the countries of western central and eastern Europe.

Chapter Six

Korea's Relations with Europe: An Overview

This chapter will examine the development of economic and political relations between the Republic of Korea and the countries of western, central and eastern Europe.[1] It is the author's belief that the increasing levels of investment in Europe since the mid-1980s reflect both a growing awareness of the importance of Europe and mounting concerns over the possible impact of European integration on ROK trade interests and export performance. This chapter also incorporates an overview of South Korean attitudes towards European integration, and assesses their impact on investment in the region.

When considering the economic, political and diplomatic relationship that exists between the Republic of Korea and Europe, perhaps the most striking aspect is the fact that it is only in the very recent past that the relationship has achieved any real measure of depth and maturity.[2] Indeed, in the case of the countries of central and eastern Europe, it is only since the late 1980s that diplomatic relations have been established with South Korea. This fact is highlighted by the paucity of materials relating to the bilateral relationship in general, and works in the English language in particular.

Although a few works were published in the latter half of the 1980s to commemorate the centenary of the establishment of diplomatic relations between Korea and various western European nations, literature searches have failed to reveal a comprehensive study of relations up to the present day, and one that includes links with the countries of central and eastern Europe. The information used in this chapter is, once again, drawn principally from Korean language sources.

KOREA AND EUROPE: A HISTORICAL PERSPECTIVE

Until the end of the 19th century, Korea's external relations were almost entirely focused on China. Korea's long and fruitful relationship with its powerful neighbour encouraged many government leaders and policy makers to concentrate their attention on relations with China and turn their back on possible contacts with the west. This isolationist policy ultimately led to a confrontation with the west similar to but on a smaller scale than that faced by China itself. China's efforts to maintain its influence over Korea following the opening up of the kingdom to the west in the late 1870s were thwarted by the rise of Japan and its eventual domination of the region. The geopolitical importance of Korea, principally as a bridgehead for a Japanese advance into China, placed it at the centre of the struggle for supremacy between Russia, China and Japan, and led ultimately to the annexation of Korea by the Japanese in 1910.

It was in the first half of the nineteenth century that western merchant ships, including vessels sent from Britain, France and Germany, appeared in numbers in Korean waters, requesting the opening of trading relations with the kingdom. However, there was little substantial contact between Korea and the major European nations until the middle of the nineteenth century, and, once diplomatic relations were established in the 1880s, there was direct contact for less than thirty years before Korea became a part of the Japanese empire. From 1910, Korea ceased to exist as an autonomous nation, and her external relations were handled by the Japanese colonial government until liberation in 1945.

The strategic importance of Korea was recognized by the Great Powers following the Allied victory in World War Two, when both the United States and Russian governments moved swiftly to establish their influence over the newly-liberated country. The outbreak of the Korean War in 1950 led to the division of the country into communist north and democratic south, and shaped the Republic of Korea's external relations for the next three decades. Political, military and economic contacts were fostered exclusively with the west and non-communist nations, and it was not until the late 1980s and early 1990s that relations were established with the eastern bloc.

In the aftermath of the Korean War and during the early years of South Korea's economic development, the Republic's external relations were focused primarily on the United States and Japan for military, political and economic reasons. At the end of the Korean War, the US government reaffirmed its commitment to the security and defence of South Korea, and assisted its early moves towards economic development through the provision of grants and loans. Following the restoration of diplomatic relations between South Korea and Japan in 1965 and the payment of reparations by the Japanese for damage inflicted during the colonial period, trading contacts developed rapidly between the two neighbours. Over the next two decades, Korea grew increasingly dependent on the United States as a major export market and on Japan as a source of imports, with both trading partners supplying capital, technology and management expertise to the Republic.

Although European governments sent troops to fight in the Korean War and diplomatic relations were established between the Republic of Korea and the founder members of the EC in 1963, the post-war years saw little substantive development in the diplomatic and economic ties between Korea and western Europe. In 1964, President Park Chung-hee visited West Germany; preparations began for the signing of trade agreements with West Germany, France and Italy, and the Republic of Korea opened a Mission to the European Community in Brussels.

According to Han Chong-su, from the Koreans' perspective, the main purpose in establishing a presence in Brussels was to secure a base from which to expand into the region. This would enable the Koreans to strengthen their relations with the EC member countries and reduce their dependence on Japan and the United States in the arenas of trade and diplomacy (Han 1998: 205). It has already been noted that, in the 1960s and 1970s, Korean companies were heavily dependent on Japan for their imports and the United States as their major export market. In view of the Republic's growing trade deficits with these two countries and the rising tide of protectionism in the USA, the Korean government began to take a keener interest in Europe, both as an alternative market for exports and as a source of imports. Nevertheless, relations between Korea and western Europe remained comparatively inactive, apart from West German President Lübke's state visit to Seoul in 1967.

The 1970s saw the beginning of the erection of non-tariff barriers in the EC as a means of protecting local industries, leading to the opening of trade negotiations between the Koreans and the EC member countries. In 1975, the ROK government concluded its first round of trade negotiations with the EC, signing an agreement on textiles and apparel which remains in force to this day. Trade restrictions on other Korean products followed, including footwear, steel, tinned mushrooms and electronics products, and by 1976, 34.6 per cent of Korean exports to the EC faced barriers of one kind or another (Han 1998: 206). At the end of 1970s, Korean exports to the EC remained at a fairly low level of US$1.9 billion, compared with US$4.4 billion in exports to the United States. In the same year, Korean imports from the EC amounted to US$1.4 billion, against US$6.0 billion in imports from Japan (Korea Foreign Trade Association 1994: 196–204).

Despite the positive support given to the ROK government by the EC member states in the United Nations and other international bodies, the focus of South Korea's diplomacy remained the United States and Japan. Furthermore, the EC still appeared to regard South Korea as a developing country, and maintained closer relations with other Asian countries such as mainland China (a trade agreement signed in 1974) and Japan (formal high-level talks from 1974 onwards).

Moreover, the suppression of human rights under the Chun Doo-hwan regime (1980–1987) caused problems in the emerging relationship; not a single high-level visit was made to South Korea by a European head of state between the assassination of President Park Chung-hee in October 1979 and the commuting of the death sentence passed on opposition leader Kim Dae-jung in 1981. (One of the key opposition figures in South Korea, Kim was sentenced to death in September 1980 for 'anti-state activities'.) Normal relations were restored in the autumn of that year, with the visit of Prime Minister Nam Duck-woo to Brussels for discussions on issues pertaining to trade and human rights (Han 1998: 208).

In the early 1980s, as Korea's rapid economic growth continued and the country matured into a Newly Industrializing Economy, relations with the EC entered a new phase. The Koreans were still eager to lessen their dependence on the United States and Japan, and the Europeans were now beginning to focus their attention on the 'economic miracle' taking place in Korea. In 1983, South Korea became the third Asian country (after Japan and mainland China) to hold annual high-level talks with the EC, covering issues in the areas of economics, trade, and scientific cooperation.

In April 1986, ROK President Chun Doo-hwan visited the United Kingdom, West Germany, France, and Belgium as part of the celebrations to mark the 100th anniversary of the establishment of diplomatic relations between Korea and Europe. The main aims of the president's visit, which was the first by a South Korean head of state for more than twenty years, were to raise the level of bilateral trade and assist Korean firms in their efforts to acquire advanced technology (Hong 1995: 89). Despite this flurry of activity, once the centennial celebrations were over, relations between Korea and the western European nations returned to their previous lacklustre state. The only developments of note in the late 1980s and early 1990s were the opening of the EC delegation in Seoul, the visit by President Roh T'ae-woo to the United Kingdom, France and West Germany in 1989, and the signing of a Science and Technology Agreement between the EC Commission and the ROK government in 1991. However, significant changes were taking place in Korea's relationship with the rest of Europe; following the end of the Cold War and the collapse of communism, the Republic of Korea softened its anti-communist stance and began to open up to its former enemies.

The turning point in relations between western Europe and Korea came in 1993 with the drafting and adoption of an EC Communiqué on EC-Korea relations. The document, entitled *Relations between the EC and Korea: Towards a Growing Partnership*, noted the need for the further development of bilateral trade, economic relations and exchanges between Europe and Korea, reflecting the importance of both sides in global trade. These and other developments indicated a change in attitude on both sides and a recognition of the mutual benefit to be derived from closer links. Cho (1993) noted the marked change in the relationship between Korea and western Europe in the early 1990s, with Korea regarding the EC as an 'increasingly valuable and reliable partner' in its efforts to diversify its trade and enhance its technological capabilities. For its part, the EC now recognized Korea's status as Asia's third largest economy and its 'potential as a serious partner in the areas of trade and economics' (Cho 1993: 74).

When it was announced that President Kim Young-sam would make a tour of major European countries in the spring of 1995 accompanied by a delegation of Korean businessmen, the reasons given for the visit revealed the sea change in South Korea's perception of and attitude towards Europe. According to the Korean press, President Kim's European visit (which included official visits to the United Kingdom, France, Germany, Denmark, Belgium, and the Czech Republic) had political and diplomatic aims too. The Kim administration was keen to widen Korea's diplomatic horizons and counterbalance the Republic's long-standing dependence on the United States and Japan in political and economic affairs. The visit also aimed to muster support for Seoul's position on reunification with North Korea, and its bid to obtain a non-permanent seat on the United Nations Security Council and membership of the Organization for Economic Cooperation and Development (OECD) (*Korea Times* 15 March 1995: 6).

However, perhaps of greatest importance in the eyes of the Koreans was their request for the establishment of a Framework Agreement for Trade and Cooperation with the EU, which would regulate bilateral relations in trade, economics, industry, science, the environment, and culture. The EU had established similar agreements with three non-European G7 members (the United States, Japan and Canada) as well as a number of Latin American and Asian governments. The EU viewed the Koreans' request in a positive light and, in March 1996, the ROK and EU initialled a Framework Agreement, calling on South Korea and the 15 member states to promote bilateral trade and cooperation in a variety of areas including science and technology, the environment, energy, telecommunications, and intellectual property rights (*Korea Newsreview* 9 March 1996: 4).

The principal aims of the agreement included the strengthening of bilateral economic relations, the establishment of cooperation in key areas (especially the fields of science, technology and industry), and the facilitation and promotion of bilateral investment. Both sides agreed to adopt measures to improve access to each other's markets, eliminate non-tariff barriers, improve the transparency of their regulations, and facilitate cooperation between the customs authorities of the EC, its member states and Korea.

The agreement also covered the area of political cooperation, calling for regular dialogue 'founded on a mutual respect for democratic principles and human rights'. Finally, the EU and ROK representatives agreed to establish a Joint Committee, comprising members of the Council of the European Union, the Commission of the European Communities and the Republic of Korea, and convening annually (alternately in Brussels and Seoul) to facilitate the implementation and further the general aims of the Agreement. The Committee would also suggest and discuss new measures for achieving these aims, monitor the Agreement, ensure its smooth running, and anticipate any future difficulties (*Baker & McKenzie European Union Newsletter* May 1996: 1). On 28 October 1996, the Framework Agreement for Trade and Cooperation was signed in Luxembourg. The agreement was subsequently ratified by South Korea and, as of March 1998, 12 of the EU member states were preparing to ratify the document.[3]

THE ASIA-EUROPE MEETING (ASEM)

Recent years have seen the creation of an additional forum for ROK-EU dialogue with the establishment of the Asia-Europe Meeting in 1996.[4] The creation of ASEM constituted an important final link in the triangle connecting Europe, Asia and the United States. Prior to 1996, issues concerning Asia and the USA could be handled under the Asia-Pacific Economic Cooperation (APEC) agreement and other Pacific Basin networks, while fora such as the World Trade Organization (WTO) and meetings of the G7 group of nations provided opportunities for dialogue between the United States and Europe. (Although there have been ministerial-level talks between the EU and the ASEAN nations since 1978, not all the nations active in ASEM have been included in this dialogue.) In creating this new arena for discussion, the Asian and European nations were acknowledging changes in the global trading order, the increasing importance of Asia-Europe multilateral ties, and the desirability of Europe assuming a larger role in the economic, political and security issues of the Asian region.

By the mid-1990s, the East Asian economies, including Korea, had become the EU's 'most dynamic trading partners', with a 26.8 per cent share of total EU imports in 1995 (up from 11.2 per cent in 1982) and a 21.3 per cent share of total EU exports (up from 8.4 per cent in the same period) (Dent 1997: 506). In June 1994, in recognition both of the economic performance being witnessed in the region and also the potential for future trade and investment, the European Commission adopted a document outlining its strategy for the future development of relations with Asia. The following year, Singapore's Prime Minister Goh Chok Tong suggested a summit meeting of leaders from Europe and Asia, focusing on issues of economic cooperation. The first gathering of the ASEM nations took place in Bangkok in April 1996, with the theme 'Towards a Common Vision for Asia and Europe'.[5]

Korean and western analysts have identified a number of factors underlying the decision to establish this forum for dialogue between Asia and Europe. Chief among these have been changes in the global economic order, including the comparative weakening of the United States' global economic and diplomatic influence, the development of a 'region-based multipolar order' (as opposed to the 'US-led multilateral order' of the past), and the creation of a new global trading environment, focusing on conflict resolution and the promotion of inter-regional cooperation (Lee Sahng-gyoun 1999: 20, 32).

For the Asian nations, closer links with Europe offered access to the advanced technology and know-how essential to their continuing economic development and growth. The summit meetings, which would afford delegates from Asia an equal status with their counterparts from Europe and signify a break from attitudes and perceptions of the past, would also provide a counterbalance to the interests of other superpowers in the region. Indeed, according to Bridges, the true significance of the creation of ASEM for the Asian nations may have lain in their wish to make it clear to the USA that recognition of their position in the global economy was growing, and they could now turn to sources of support other than their 'sometimes overbearing' relationship with the United States (Bridges 1999: 182–183).

For the Europeans, a similar motivation could be observed in that they shared concerns over the political and economic links already established between the United States and Asia, and had reservations regarding the domination of APEC by the USA. (The EU has been denied observer status at the APEC meetings.) It has been argued that part of the rationale for creating ASEM was the prospect that such an organization might 'check US impulses towards neo-isolationism and unilateralism' (Maitland and Yu 1998: 16). In addition to the possibility of gaining a degree of influence over political and security developments in Asia, ASEM offered the Europeans a chance to secure a strategic position in a region with substantial potential as a market for European trade and investment.

For the Koreans, ASEM also offered the chance to pursue a policy of multilateral diplomacy, moving away from its traditional economic, political and diplomatic dependence on the United States and Japan (Han Sung-joo 1996: 146). Another important factor for the Koreans was the issue of relations with North Korea. In contrast to the United States, which began moving towards the establishment of diplomatic and economic relations with the DPRK in the mid-1990s (See Chapter 7), the Europeans have, in general, remained reluctant to develop a relationship with North Korea. Direct approaches from the North Koreans, asking for economic or commercial contacts, have not met with success and the European Commission has consistently sought and abided by the opinions of the ROK government on inter-Korean affairs (Nuttall 1998: 180).

Bridges maintains that one of the reasons why the South Koreans attached such great importance to the drafting and ratification of the Framework Agreement mentioned above was that the document was perceived to support the ROK government's position on relations with North Korea (Bridges 1999: 63). Although the United States has offered reassurances that the development of contact with North Korea will not occur at the expense of ROK-US relations, it must have appeared to the South Koreans that the development of closer ties with Europe through ASEM would afford them a greater degree of international support in their efforts to achieve the peaceful reunification of the Korean peninsula.

Discussions at ASEM 1 covered a broad range of topics of mutual interest, including trade and investment, science and technology, security, energy, transportation, health, environment, crime, and cultural links.[6] Delegates to the meeting recognized their 'common commitment to a market economy, open multilateral trading system, non-discriminatory liberalization, and open regionalism', and viewed the initiatives launched at ASEM as complementing and reinforcing, rather than conflicting with, the work of other international bodies, such as the WTO (Asia-Europe Meeting 1 1999: 157–159). The Chairman's Statement at the end of the summit noted that:

> The Meeting recognized the great potential for synergy between Asia and Europe on account of the economic dynamism and diversity of the two regions. Opportunities exist for both regions to expand the market for goods, capital equipment and infrastructure development projects, and to increase the flows of capital, expertise and technology (Asia-Europe Meeting 1 1999: 157).

ROK-EU ECONOMIC RELATIONS

Although the past three decades have seen encouraging rates of growth in trade and investment between the Republic of Korea and the countries of western Europe, economic exchanges remain well below the levels seen between Korea, Japan and the United States. Korea has enjoyed certain privileges in its trade with the EU, but friction has been caused by the EU's imposition of import controls and the ROK government's restrictions on access to the Korean market.

Trade

It will be remembered that, in the late 1970s, the absolute level of bilateral trade between Korea and the EC was low in comparison with exports to the United States and imports from Japan. Growth rates in the 1980s and 1990s, however, were impressive, according to statistics published by the Korea Foreign Trade Association (KFTA). Korea's exports to the EC more than trebled between 1980 and 1993, rising from under US$3 billion to more than US$9 billion, while there was a five-fold increase in the Republic's total exports. During the same period, imports to Korea from the EC jumped from under US$2 billion to break through the US$10 billion mark, while the

EC's total imports almost quadrupled. The rate of growth of trade between the ROK and the EC outpaced that of Korea and the United States or Japan for the same period (Kim Bak-soo 1994: 3).

Although the 1980s and early 1990s saw a remarkable increase in Korea's exports to the EC, the Republic's share of the EC market remained small, standing at just 1.5 per cent of imports to the Community and 1.4 per cent of exports from the EC. Furthermore, the share of Korean exports accounted for by the EC was declining: in 1980, the EC was the destination for 15.5 per cent of Korean exports but this share had fallen to 11.4 per cent by 1993. The decline in Korean exports to the EC was due in large part to the loss of price competitiveness following substantial wage rises at home, and increasing competition from China and South-east Asian countries with lower wage costs and growing manufacturing bases. In contrast, between 1980 and 1993, the EC's share of the South Korean market had increased from 7.2 per cent to 12.1 per cent (Kim Bak-soo 1994: 3).

As can be seen in Table 6.1, ROK-EC trade shrank by 4.4 per cent in 1992, the first year-on-year decline since the establishment of the EC. The drop in trade volume was due to the economic slowdown experienced by both partners, and the increase in EC imports from eastern Europe and emerging economies such as mainland China. However, the downward trend was reversed in 1993, and trade continued to expand over the next three years, showing particularly strong growth in 1994 and 1995, when bilateral trade increased by 22.4 per cent and 33.8 per cent respectively, year on year. Within the space of five years (between 1991 and 1996) the volume of trade had jumped from slightly more than US$21 billion to over US$36 billion. The balance of trade, which had been in Korea's favour throughout the 1970s and 1980s, turned towards the EU in 1991, after which the deficit steadily expanded, reaching US$5.9 billion in 1996 (Lee Chong-hwa 1997: 9).

In 1996, Korean imports from the European Union amounted to US$21.2 billion (a rise of 16.6 per cent from 1995) and accounted for 14.1 per cent of all Korean imports. In the same year, the United States was the largest supplier to Korea with a 22.2 per cent share, followed by Japan with 20.9 per cent. The USA was also the major importer of Korean products, with a 16.7 per cent share of Korean exports, followed by Japan with 12.2 per cent. Korean exports to the EU increased by 6.0 per cent year on year in 1996 to reach US$15.3 billion, or 11.8 per cent of Korea's total exports in that year. In 1996, Korea was the EU's ninth largest trading partner, while the EU stood behind the USA and Japan as Korea's third largest trading partner (Oh Jong-nam 1997: 281).

Trade privileges

The Treaty of Rome (1957) provided for cooperation between the EC and developing countries with which it had special relationships; these countries were headed by former European colonies, which were grouped together under the Lomé Convention (1975). Being neither a former colony nor classified in the same group of countries as India, Pakistan or the ASEAN nations, with which the Community had concluded cooperation agreements, the Republic of Korea did not qualify for a special relationship or form of association with the EC.

In 1971, the EC introduced the General System of Preferences (GSP), under which developing countries including the Republic of Korea could benefit from non-reciprocal customs duty exemptions for manufactured and non-manufactured goods, and some processed agricultural products. However, having accorded these trade privileges to South Korea for more than two decades, the EC announced in 1994 that 50 per cent of GSP benefits for a number of industrial products manufactured and exported by Korea would be withdrawn the following year. Included in the rationale behind the decision was the view that, as South Korea had announced its intention to apply for membership of the OECD in 1996, it would be inappropriate for the Republic to receive benefits afforded to developing countries.

In 1993, it was estimated that 21 per cent of ROK exports to the region received GSP benefits (Han 1998: 219); according to Korean analysts, the loss of GSP benefits could result in an annual decline of around US$2 billion in the Republic's exports (*Korea Economic Weekly* 13 February 1995: 2). In January 1996, GSP benefits were withdrawn from some products including major exports items such as automobiles, consumer electronics products, footwear, textiles and apparel; the remaining benefits were withdrawn in May 1998 (Lee Chong-hwa 1997: 43).

Trade friction and restraints

Trading relations between the EU and non-member countries are currently covered by the Common Commercial Policy, which governs external customs tariffs, the conclusion of tariff and trade agreements, the alignment of liberalization policies and measures among member countries, export policy, and protective measures. In addition to the common policy adopted by the EU, individual member countries can take steps to prevent the entry into their own market of specific goods that are circulating in other member country markets. Such restrictions include anti-dumping investigations, rules of origin regulations, and screwdriver plant regulations.

Anti-dumping investigations are carried out when it is suspected that foreign goods are being dumped in the market: that is, they are being sold at a lower price than in the home market. An investigation may also be carried out when it is suspected that imports pose a threat of 'material injury' to a country's industry.

These investigations may lead to the imposition of duties, the adoption of price undertakings by the exporter, or a ruling that dumping has not, in fact,

Table 6.1 ROK-EU Trade, 1990–1996

	1990 US$bn	*1991 US$bn*	*1992 US$bn*	*1993 US$bn*	*1994 US$bn*	*1995 US$bn*	*1996 US$bn*
ROK Exports to EU	10.0	10.6	9.8	9.9	11.3	16.3	15.3
YOY (%)	22.1	5.0	(6.7)	1.0	13.2	44.9	(6.0)
EU Exports to ROK	9.1	10.7	10.5	11.1	14.5	18.2	21.2
YOY (%)	28.9	17.9	(2.1)	6.1	30.6	25.1	16.6
Total	19.1	21.3	20.3	21.0	25.8	34.5	36.5
YOY (%)		11.1	(4.4)	3.7	22.4	33.8	5.9
ROK Trade Balance	0.9	(0.1)	(0.7)	(1.2)	(3.2)	(1.9)	(5.9)

YOY Year-on-year growth rate
Source: Lee Chong-hwa 1997:9.

taken place. Rules of origin regulations are applied in order to determine the country of origin of a given product and thus decide whether or not duties should be imposed. The country of origin is generally taken to be the country in which the 'last significant process' in the manufacture of the product took place. The screwdriver plant regulation, adopted in 1987, aimed to ensure that foreign firms did not evade anti-dumping duties by performing simple assembly operations of imported parts and materials within Europe. The regulation required non-member manufacturers to source at least 40 per cent (in value terms) of the parts and materials used to assemble a given item from outside their own country (Byun 1995: 135–140 and Shin 1998: 25).

As the competitiveness of products manufactured in Europe weakened in the 1970s (See The Korean View of European Integration), Korean exporters were able to expand their presence within the European market with high-volume, low-priced exports. Within a short space of time, however, the Koreans found themselves facing trade barriers in the European market, as trading partners took action to stem the flow of cheap imports. The restrictions have continued to the present day.

In 1993, Korean products faced 60 import restrictions in industrialized countries, including 19 in the United States, 16 in Europe, and eight in Japan. Just as the total number of restrictions has fallen in recent years (from 77 in 1989), the number of cases recorded in Europe has dropped from 28 in 1989. Of the 16 restrictions in place in 1993, four took the form of quota restrictions and 12 were anti-dumping duties or other price restraints. The majority (13 restrictions) were imposed in the region as a whole, with only three being imposed in individual member countries. In line with the decline in the number of restrictions, the proportion of Korean exports subject to restrictions has fallen from 31.6 per cent in 1976 to 30.0 per cent in 1985, 29.5 per cent in 1990, and 24.9 per cent in 1993 (Han 1998: 217–218).

Cho Wŏn-dong has argued that Korean products have been more heavily targeted than those of other nations in recent years. Quoting KFTA statistics, he showed that 26.3 per cent of Korean exports to the EC in 1992 were subject to anti-dumping duties, compared with 16.0 per cent of Korean products in the United States, 17.4 per cent in Canada, 8.1 per cent in Australia and 7.4 per cent in Japan (Cho 1993: 86). It is certainly true that the structure of ROK exports to the EC made Korean manufacturers a target for protectionism in that trade was focused on a small number of items in a few markets. The Koreans targeted as strategic export sectors those industries such as consumer electronics in which they faced heavy competition but enjoyed only a price advantage *vis-à-vis* their competitors. The decision to achieve economies of scale by mass producing and exporting the output to a small number of markets meant that ROK exports to the EC had a significant impact on specific industries and areas of Europe, resulting in pressure from local producers to take action against the inflow of Korean goods.

For their part, the Europeans have been critical of the ROK government's trade policies and the restrictions placed on access to the Korean market for foreign goods and services. Among the major trade issues discussed by the South Koreans and Europeans are the protection of intellectual property rights (IPR), access to South Korea's alcoholic beverages, automobile and financial services markets, and the procurement of telecommunications equipment. In addition to item-specific trade issues, there have been more general problems, such as the anti-import campaigns periodically carried out in Korea and the imposition of 'adjustment tariffs' which were being applied to around 10 per cent of imports into South Korea in the early 1990s (Lee Chong-hwa 1997: 25).

EU investment in Korea

Investment in South Korea by foreign companies began in 1962, with Hoechst becoming the first European investor in the Republic two years later. Over the next three decades, a total of US$10.2 billion was invested by foreign firms; Japanese investment accounted for 41 per cent of the total, American for 29 per cent, and European for 19 per cent. Although South Korea's major trading partners, the United States and Japan, together accounted for 70 per cent of the cumulative total in 1992, the share of FDI accounted for by European firms has been increasing since the latter half of the 1980s.

Between 1962 and 1971, the United States and Japan accounted for 82 per cent of inward investment, with the Europeans taking a share of just 8 per cent. Between 1972 and 1986, the shares were 82 per cent and 7 per cent but over the next five years, the balance shifted with the USA and Japan taking a smaller share of 64 per cent, while the share accounted for by European firms increased to 26 per cent. The larger share reflected the greater level of interest in the South Korean market on the part of the Europeans, and

Table 6.2 Investment in Korea by Major Western European Nations, 1962–1996

	1962–88 US$mm	*1989–91 US$mm*	*1992 US$mm*	*1993 US$mm*	*1994 US$mm*	*1995 US$mm*	*1996 US$mm*	*1962–96 US$mm (%)*
Netherlands	148.0	655.0	44.0	131.0	67.0	170.0	205.0	1,420.0 (31.0)
Eire	0.0	0.0	0.0	1.0	164.0	73.0	410.0	649.0 (14.2)
France	105.0	91.0	29.0	40.0	56.0	35.0	90.0	447.0 (9.8)
Germany	185.0	179.0	120.0	36.0	60.0	45.0	95.0	721.0 (15.7)
UK	135.0	110.0	24.0	71.0	25.0	87.0	79.0	531.0 (11.6)
Switzerland	177.0	140.0	37.0	7.0	11.0	10.0	162.0	543.0 (11.7)
Others	55.0	73.0	28.0	22.0	23.0	56.0	17.0	275.0 (6.0)
Total	806.0	1,248.0	282.0	307.0	407.0	475.0	1,058.0	4,584.0 (100.0)

Source: Cheong 1997: 169.

intensified efforts on the part of the ROK government to diversify its investment and technology sources away from the USA and Japan (Han 1998: 220).

As can be seen in Table 6.2, as of the end of 1996, the cumulative amount of investment in South Korea by EU member nations stood at US$4.6 billion. Of the total, US$3.8 billion or 82.4 per cent had been invested since 1989. In the period between 1962 and 1996, the Netherlands was the largest investor in South Korea with 31.0 per cent of the total invested; the Netherlands was followed by Germany (15.7 per cent), Eire (14.2 per cent), Switzerland (11.7 per cent), the United Kingdom (11.6 per cent), and France (9.8 per cent).

The Korean view of European integration

Although the main focus of South Korean political and economic interest has traditionally been the United States and Japan, this does not mean that ROK government officials, manufacturers and exporters have failed to take an interest in developments in Europe. Moves towards a single European market and full economic, monetary and political integration created both concern among Koreans over the prospect of 'Fortress Europe' and interest in the opportunities that economic integration might offer. This section examines the attitudes of Korean scholars towards the Single European

Market in the late 1980s and early 1990s, providing an insight into some of the motivations for FDI in western Europe in the first half of the 1990s.

Following the two oil shocks, the EC economy entered a period of stagnation, suffering from low economic growth, high inflation and high unemployment. EC member countries were lagging behind other industrialized nations including the United States and Japan in key high-tech sectors such as computers. The EC was also facing new challenges from the Asian Newly Industrializing Economies (NIEs, including the Republic of Korea) in traditional sectors such as textiles and shipbuilding, where the competitiveness of European firms was declining.

A number of reasons for this loss of competitiveness have been identified by Korean scholars. According to An *et al*, the growth rate for exports from the EC to non-member countries was slower than the region's overall export growth rate throughout the 1960s and 1970s. This was because European companies were focusing their attention on local sales rather than on competing in overseas markets. This heavy dependence on and success in regional trade led to a lack of motivation for the companies to boost their international competitiveness. Another side effect was a neglect of the industrial restructuring that would be needed to meet the coming challenge from the NIEs (An *et al* 1991: 17–21).

European investment in research and development (R&D) in high-tech fields lagged behind levels seen in the United States and Japan, and the stagnation of corporate investment in new products and R&D led to delays in restructuring, resulting in a further decline in competitiveness (An *et al* 1991: 21–24). This 'Eurosclerosis' prompted the governments of the EC member countries to promote the reactivation of the European economy through industrial restructuring. The policy adopted by the EC combined two basic strategies: they would protect 'sunset' or declining sectors while nurturing new strategic industries and enhancing member countries' competitiveness, with the ultimate aim of full market integration. These moves towards the creation of a single market in the 1980s triggered fears of 'Fortress Europe' among the Koreans; those concerns intensified as western Europe began to move towards full economic integration in the early 1990s.

The debate – 'Fortress Europe' or 'Opportunity Europe'?

It is generally accepted that the integration of the European market to form the world's largest single trading bloc was a major factor in the surge of Korean interest in Europe in the 1990s. Integration was viewed by Korean scholars and commentators from two contradictory viewpoints in the late 1980s and early 1990s. The pessimists held that European integration would create a barrier to the Republic's export success, as protectionism increased and outsiders were locked out of 'Fortress Europe'. The more optimistic view of events in Europe was that many of the opportunities and benefits of integration enjoyed by member countries would be shared by non-members too.

The Koreans' negative view of European integration reflected concerns regarding the phenomenon of regionalism, or the formation of world economic blocs. According to Chŏng and Pak, the fundamental aim of regionalism was to seek to secure the benefits of free trade by removing tariffs and other trade barriers within an expanded market. However, the trend towards regionalism could only be said to be a desirable phenomenon if internal and external trade and economic cooperation became more brisk as a result of integration. The concerns expressed over the formation of trading blocs focused on the anticipated increase in protectionism towards non-member countries and efforts to restrict the benefits of integration to companies within the bloc. Such developments would make it essential for an export-dependent country like South Korea that did not belong to an established regional bloc to develop an appropriate response to the phenomenon (Chŏng and Pak 1992: 18).

'Fortress Europe'

The negative view of the potential impact of European integration on Korean economic interests was most clearly expressed by Chŏng and Pak. They argued that, given that the aim of economic integration was to secure benefits for member countries, it was inevitable that a policy of selective protectionism would be implemented *vis-à-vis* firms from non-member countries which enjoyed a competitive advantage (Chŏng and Pak 1992: 21). Ŏm claimed that, although the EC advocated free trade in its external trade policies, it had, in fact, been strengthening its protectionist policies. This action was prompted by concerns regarding the region's chronic financial deficits, the delay in industrial restructuring, the increase in structural unemployment, the lag in technology *vis-à-vis* the United States and Japan, and the increase in pressure from political pressure groups such as unions and industry associations to protect declining industries (Ŏm 1991: 73).

In addition to general fears that integration would bring heightened protectionism against non-member countries, there were also more specific concerns over the possible impact of trade diversion, the increased competitiveness of European firms, non-tariff barriers, standards harmonization, reciprocity, and the further development of relations with eastern Europe.

Kim Bak-soo noted that, in the past, growth in EC trade had been driven by the expansion of internal trade among member nations. Should that trend continue, he argued, the trade diversion effect of integration would far outweigh any trade creation effect secured by the formation of a single European market. Moreover, unless European production lagged behind demand, companies from non-member countries might experience difficulty increasing their trade volume with the EU, as member countries boosted intra-bloc trade by switching their import sources from companies outside the region to firms from EU member states (Kim Bak-soo 1993: 104).

The main effects of integration in terms of the increased competitiveness of European firms were summarized by An *et al.* As physical, technical, legal and fiscal barriers within the bloc were removed, the reduction in production costs would lead to a fall in product prices. Demand would increase and production would expand, leading to a further reduction in costs and an increase in technology innovation. This would contribute to a strengthening of corporate competitiveness, which would also be boosted by economies of scale and the learning effect. The restructuring of European companies and industry would also be promoted, as business groups sold off inefficient firms and companies developed strategic alliances to develop new technology and pool R&D resources (An *et al* 1991: 36).

A country like South Korea, which had in the past enjoyed a competitive edge in terms of price rather than quality, would face the erosion of its competitive advantage, and this would prove hard to recover in view of the upward pressure on production prices back at home. Even if external trade barriers were not strengthened, South Korean firms might find it harder to penetrate the European market or expand their market share, given their weak competitiveness in terms of cost, brand image and marketing strategies.

Among the trade regulations and non-tariff barriers imposed on ROK imports to Europe, anti-dumping duties and local content or country of origin regulations caused the most concern. The rapid increase in the number of anti-dumping investigations against major Korean products, compared with other competitor nations, was cited as evidence of the 'arbitrary and discriminatory nature' of the regulations (Min 1991a: 164). The definition of the 'country of origin' as that in which the 'last significant process' takes place had made it more difficult for a foreign company's products to qualify as European goods. This was because a product would only be deemed to be European if nearly all the important and costly stages of the production process were carried out in Europe. Products that did not meet the local content and country of origin requirements could face import restrictions or anti-dumping duties. In connection with this development, Min expressed concern that requiring firms from non-member countries to source key parts and components within Europe might leave them vulnerable to difficulties in sourcing them at competitive prices from European competitor companies (Min 1991b: 57).

The harmonization of technical standards throughout the EU was expected to result in economies of scale and a reduction in costs of compliance for member and non-member countries alike. All companies selling products in the EU would have to conform to essential requirements relating to health, safety, consumer and environmental protection, as well as technical standards determined by organizations including CEN (Comité Européen de Normalisation) and CENELEC (Comité Européen de Normalisation Électrotechnique). However, once the standards were met, the products could be sold in any other country within the region without additional testing and certification. Should a manufacturer produce goods in accordance with standards other than those adopted by the EU, he would have to obtain separate approval, incurring additional expense.

This process of harmonization raised fears that the new technical standards could be used as a barrier against non-EU countries. Should the new standards be set at a level with which a non-member country could not comply, the standards would effectively become a new technical barrier and that country's products would be shut out of the market (Son and Yi 1991: 164). There were also concerns that, even if South Korean products did meet the standards, the stringent testing and certification procedures required for imported goods and delays in testing could result in losses for ROK exporters. In addition, as non-member countries were excluded from the process of defining standards, which in itself was not completely transparent, there was concern that the standardization process could be used to block non-member countries' entry to the EU in sectors where they had a competitive edge. This could be achieved by setting higher and more stringent standards for goods of which imports were increasing (Chŏng and Pak 1992: 27).

Although integration would create opportunities for non-member countries, it had been made clear that these benefits could not be enjoyed without some degree of reciprocity: that is, non-member countries would have to afford to the EU the same privileges and market access as they were seeking from member countries. The Europeans were particularly keen to apply the principle of reciprocity in the areas of finance, telecommunications, distribution and other services, public procurement, and mergers and acquisitions (M&A).

In 1990, Kim Ji-hong expressed concern that demands for reciprocity by sector would work against the principles of multilateral trade liberalization, and that any resulting discrimination would, in all probability, be targeted at a number of countries, including Korea (Kim Ji-hong 1990: 234). Paek Chin-hyŏn concurred that the demand for reciprocity as a condition for entry to the European market caused the most worry to non-member countries. Paek noted that, should 'reciprocity' mean that non-member countries must guarantee treatment equal to that which they receive in Europe, the application of reciprocity could greatly restrict their entry into the region in certain sectors (Paek 1991: 163). The requirement of reciprocity in M&A opportunities, for example, could restrict penetration of the European market by Korean companies, if the ROK government were unwilling to allow takeovers of domestic corporations by EU firms.

Fears over the impact of economic issues were compounded by concerns about the implications of the further expansion of the EU, from 15 members in 1995 to include the EFTA nations and a number of eastern European nations (including Hungary, Poland, the Czech Republic and Slovakia). Given the increasing possibility of the formation of a 'pan-European bloc'[7], Ŏm expressed the view that the expansion of agreements on trade privileges between the EU and its neighbours would enhance the competitiveness of non-member, allied countries in the European market. Moreover, he noted that, in the event of the EU continuing to expand trading privileges and cooperation with surrounding countries, non-cooperating countries such as Korea would be placed in an even more disadvantageous position when trading with Europe (Ŏm 1991: 76, 107).

'Opportunity Europe'

In the eyes of Koreans who viewed integration as an opportunity rather than a threat, the single European market promised to be a 'remarkable playing field for outside operators' (Park Gi-han 1995: 37). The main benefits of European integration for non-member countries would be the creation of the world's largest trading bloc with 328 million consumers and cost reductions deriving from the removal of physical, technical, legal, fiscal and procurement barriers. Intra-regional trade would be facilitated by the free movement of goods, capital, personnel and services, and trade with the EU would be boosted (Paek 1991: 161). Furthermore, it would no longer be necessary for companies from non-member countries to have their goods tested and certified for sale and distribution in each member country. Other benefits derived would include increased access to public procurement markets and the potential for gaining access to the EU market by advancing into countries currently waiting to become members.

It was forecast that the formation of a single market would result in a 4.5 per cent increase in GDP, a 6.2 per cent fall in consumer prices, and the creation of almost two million new jobs (Paek 1991: 161). The fall in production costs and product prices after integration was expected to stimulate demand and production, creating employment, boosting disposable income, and further stimulating trade (Shin and Yi 1993: 22). The acceleration of economic growth in Europe would result in greater demand for commodities from the rest of the world (assuming that European production lagged behind demand), and the benefits would be compounded as the region continued to integrate with its non-member neighbours and the market expanded further. Korean firms operating in a broad spectrum of industries, from high-tech products to consumer goods, would be able to exploit new opportunities offered by the single market with rising levels of disposable income (Lee Hong-gue 1990: 262).

The integration of the market and the removal of physical, legal, financial and technical barriers would result in lower production costs and enhanced productivity. Cost reductions would be achieved through economies of scale, more efficient distribution, and the abolition of administration including border controls and documentation. The removal of border controls among EU member countries and the deregulation of the European transportation market would allow non-member country firms to benefit from further cost reductions. It would also enable them to create a more efficient distribution network through the establishment of a centralized distribution site capable of promptly serving all member states (Chŏng and Pak 1992: 18–19).

As a result of the harmonization of standards within the EU, products would, in the future, only have to comply with one standard instead of a dozen or more, as in the past. In common with member country companies, therefore, Korean firms would benefit from the elimination of the direct and indirect costs arising from compliance with differing standards and regulations, and would also be able to supply one product to the European market rather than different products to individual markets. The removal of technical barriers through the harmonization of standards would also, it was argued, result in the creation of strategic alliances and partnerships, and progress in technology and technical development (Shin and Yi 1993: 22). This development would make collaborative research more feasible, offering the Koreans opportunities to become involved in research programmes with leading European companies.

The last of the economic benefits of integration forecast by Korean commentators was the opening of the European markets for public procurement, which would give foreign suppliers better access as far as non-excluded sectors were concerned. (The four excluded sectors were energy, transport, water and telecommunications.)

As was seen earlier in the 'Fortress Europe' debate, an extra dimension was added by the proposed expansion of the EU to include former eastern bloc nations. Due to its strong anti-communist stance following the Korean War, South Korea had little contact with eastern Europe before 1989, when diplomatic relations with Hungary were established. Stronger relations between the EU and the eastern European countries would also benefit the Koreans, who would have the chance to acquire more information about and experience of working in the region. In the medium- to long-term, Korean firms would be able to take over or invest in companies in eastern Europe, thus establishing a low-cost production base from which to penetrate the EU market (Jung 1991: 170).

Finally, as mentioned above, the early 1990s saw the development of a new dimension in ROK-European political relations. As the Republic sought to have a wider role and a louder voice in international affairs and the formulation of international policy, it would become more desirable for South Korea to have close relations with the powerful European Union. In additional to the financial benefits to be gained from integration, the establishment of closer political ties with Europe could help South Korea achieve its goals of globalization and internationalization.

In conclusion, the prevailing view among Korean scholars regarding the likely impact of European integration on Korea's trade and investment interests in the early 1990s was a pessimistic one. The main concerns about the formation of a single market centred on the prospect of a strengthening of protectionist policies, either as a means of protecting declining industries during the process of industrial restructuring or with the aim of reserving the benefits of integration for member country firms. It is interesting to note that the mood seemed no more optimistic by the end of the decade; in Nam and Slater's 1997 survey of Korean investors in Europe, more than half the respondents expected full integration to have a negative impact on their business, while only 16 per cent thought Korea would benefit from the creation of a single market (Nam and Slater 1997: 45).

SOUTH KOREA'S RELATIONS WITH EASTERN EUROPE

Although diplomatic relations were established between Korea and the Austro-Hungarian Empire in 1882, they were severed when Korea came under Japanese colonial rule in 1910. From the end of the Korean War in 1953 until the late 1980s, successive governments of the Republic of Korea maintained a strong anti-communist stance in their external relations. During this period, there were no official diplomatic links between South Korea and communist states, and as late as 1990 it was impossible to travel directly from Seoul to the former Soviet Union or the Chinese mainland. Despite the presence of small Korean ethnic communities in the Soviet bloc, there were no diplomatic or commercial ties with eastern European countries until the late 1980s.

The improvement of relations between South Korea and eastern Europe was a direct result of the 'Northern Diplomacy' adopted towards the end of the 1980s by the Roh T'ae-woo administration. As the cold war era came to a close and the collapse of communism accelerated, the ROK government moved to establish formal diplomatic relations with a number of countries in the former communist bloc. The moves were surely motivated as much by economic pragmatism as by political realities, as the end of the Cold War era brought significant new trading and investment opportunities for the South Koreans.

Following the establishment of a trade office in Budapest in 1987, the Republic of Korea moved to establish full diplomatic relations with Hungary two years later. In less than a year (between February 1989 and March 1990), South Korea also established diplomatic relations with Poland, Czechoslovakia, Romania and Bulgaria. Political ties soon led to the signing of a variety of economic agreements covering trade, investment, and cooperation in the areas of science and technology.

The agreements were supported by a number of state visits to South Korea, including those by President Havel of Czechoslovakia in 1992 and President Walesa of Poland in 1994. The following year, President Kim Young-sam became the first South Korean president to make a state visit to an eastern European country, by including the Czech Republic in his tour of Europe. Common themes of discussions that took place during these visits were the expansion of bilateral trade and investment, cooperation in science and technology, and joint entry into third country markets to capitalize on South Korea's capital and technology and eastern Europe's low-cost labour.

In 1992, an ROK-Hungary Science and Technology Committee (*Kwahak Kisul Kongdong Wiwŏnhoe*) was established to undertake research projects in areas such as computers, resource management and agriculture. Three years later, cooperation agreements for science and technology and nuclear power technology were signed with the Czech Republic. Economic cooperation between South Korea and eastern European countries has also included the provision of aid, grants and assistance by the ROK government to its new trading partners. As of 1994, 45 trainees had been invited to visit South Korea, while four specialists had been dispatched to eastern Europe, and equipment and materials valued at almost US$500,000 had been sent to major eastern European countries.

The Economic Development Cooperation Fund (ECDF – *Kyŏngje Kaebal Hyŏmnyŏk Kigŭm*) was established by the ROK government in the early 1990s; among its activities has been the provision of funds to support the modernization of the telecommunications systems in Poland and Romania (US$50 million to each country, repayable over 20 years). Other forms of support from the ROK government remained at low levels in the mid-1990s; at the end of 1994, there were 19 cases of export credit totalling US$120 million, and one case of FDI support, for a joint venture in Romania (Kim Kyu-p'an 1995: 9–10).

Trade with eastern Europe

In the five-year period between 1991 and 1996, South Korea's bilateral trade with eastern Europe increased from US$2.2 billion to US$5.0 billion, comprising US$3.2 billion in exports to and US$1.8 billion in imports from eastern Europe. However, despite the rapid growth in absolute terms during that period, trade with eastern Europe accounted for a relatively small proportion of South Korea's external trade (between 0.6 and 2.4 per cent). The scale of trade has been limited by the small size of the market in many eastern European countries, the weak competitiveness of their products, and the economic downturn in the region in 1992 and 1993, which led to a stagnation of exports from eastern Europe and a strengthening of restrictions on goods into the region (Kim Sang Kyom 1997: 48–49) (See Table 6.3).

Table 6.3 Trends in Bilateral Trade, South Korea and Eastern Europe, 1990–1996

	1991	*1992*	*1993*	*1994*	*1995*	*1996*
Exports US$mm	1,344	493	948	1,360	2,066	3,168
YOY %		(63.3)	92.3	43.5	51.9	53.3
%TX	1.9	0.6	1.2	1.4	1.6	2.4
Imports US$mm	896	319	1,093	1,468	1,911	1,807
YOY %		(64.4)	242.6	34.3	30.2	(5.4)
%TM	1.1	0.4	1.3	1.4	1.4	1.2

YOY Year-on-year growth rate
%TX Percentage of total exports
%TM Percentage of total imports

Source: Kim Sang Kyom 1997: 48–49.

In 1994, Poland was South Korea's largest trading partner in eastern Europe, accounting for 47.6 per cent of exports to and 39.9 per cent of imports from the region. Poland, the Czech Republic, and Hungary together accounted for about 70 per cent of trade with the region in 1994. South Korea's exports to eastern Europe have been dominated by electrical and electronics goods, textiles, machinery, and transportation equipment, which together accounted for almost 90 per cent of ROK exports to the region in 1994. Imports from eastern Europe have, on the other hand, focused on steel, chemicals, and textiles (Kim Kyu-p'an 1995: 13–15).

CONCLUSIONS

It is no exaggeration to say that it is only in the past fifteen years that Korea has established strong and meaningful relationships with the countries of western, central and eastern Europe. Although some efforts were made by the Koreans to develop closer ties with the EC in the 1970s and 1980s, signs of real interest in the bilateral relationship only appeared in the latter half of the 1980s.

In the past decade, the Koreans have come to realize the importance of Europe, as a counterbalance to the Republic's traditional dependence on the United States and Japan, as a major market for Korean exports and host for Korean FDI, as a provider of technology and expertise, and as a supporter of Korea's endeavours to consolidate her place on the world stage. International developments, such as the establishment of ASEM, which provided the final link in the Asia-United States-Europe triangle, have reinforced the perception that Europe is of crucial importance to the Koreans.

An analysis of these developments leads the author to believe that a key factor in the surge of interest in western Europe was the creation of a Single European Market: the largest single trading bloc in the world. The realization of the importance of this development, whether in terms of the threat to Korea's interests or the opportunities offered to exporters and investors, came at a time when the full impact of import restrictions and non-tariff barriers was beginning to be felt. For the Koreans, being locked out of 'Fortress Europe' was not an option; in addition to the market potential for Korean goods, the countries of the EU offered access to the technology and expertise that Korea would need in the process of industrial restructuring and globalization.

The establishment of diplomatic relations with the countries of eastern and central Europe is also a recent phenomenon; few of these countries can claim diplomatic or trading links with South Korea with a history of more than ten years. For the Koreans, the importance of these countries lies in their medium- to long-term export market potential, their attraction as a low-cost manufacturing base, and their geographical proximity to the industrialized markets of the European Union.

This assessment of the reasons underlying the strengthening of economic links between Korea and the countries of western, eastern and central Europe over the past decades provides part of the context within which patterns of Korean foreign direct investment in Europe can be understood. The next chapter will examine Korean investment in Europe within the context of this evolving relationship, focusing on investments made by the major consumer electronics companies.

Chapter Seven

Korean Foreign Direct Investment in Europe

A report published by the United Nations Conference on Trade and Development (UNCTAD) in 1997 noted that the European Union hosts almost 40 per cent of global investment flows. Notwithstanding this leading role in worldwide FDI, the flows between Europe and Asia have been small:

> Asian investors have paid relatively little attention to the EU and other parts of the European continent. While it may be appropriate to describe European firms as neglecting Asia, Asian firms are beginning to discover Europe. European firms have failed so far to build upon a long-standing base in Asia; Asian firms are just at the beginning of their entry into Europe (UNCTAD 1997: iii–xiv).

OVERVIEW

Korean investment in Europe began in 1974, when the Bank of Korea gave approval to the Daewoo Group to establish two subsidiaries: Daewoo UK Ltd, a general trading company dealing in textiles, apparel and miscellaneous goods in London, and Daewoo Handels GmbH, trading electronics products, textiles and footwear in Frankfurt. This early investment was followed by Samsung Co. in Germany (1975), Kukje and Daewoo in France, Lucky Goldstar, Sunkyong and Hyosung in Germany, and Samsung Co. and Sunkyong in the United Kingdom (all in 1976) (Chŏng 1995: 108).

In the first half of the 1970s, Korean FDI in Europe was spearheaded by the General Trading Companies (the trading arms of the *chaebŏl*), and it was not until the early 1980s that investment began in the manufacturing sector. Between 1980 and 1984, three manufacturing projects were undertaken by Korean firms in Europe: Hanbi Sanŏp (footwear) in the Netherlands in 1980, Hŭngmyŏng Kongŏp (containers) in London in 1982, and Samsung Electronics (electronic and electrical goods) in Portugal in 1982 (Yu 1994: 26).

During the 1970s and early 1980s, investment by South Korean corporations in Europe remained at levels far below those seen in South-east Asia and North America. These low levels of investment reflected the fact that both Korea and Europe were focusing their attention on other markets, as discussed earlier. In addition, until relatively recently, the Koreans still looked on Europe as a sales market rather than a manufacturing base; investment was, therefore, centred on small-scale trade-oriented projects rather than capital-intensive manufacturing plants (Chŏng 1995: 108).

Despite this slow start, the past fifteen years have seen rapid increases in the amount invested in Europe by Korean firms, reflecting growing concerns over the strengthening of trade restrictions, the formation of a Single European Market and the possible creation of a 'Fortress Europe', as well as the recognition of the business prospects presented by the opening up of central and eastern Europe. As can be seen in Table 7.1, having remained well below the US$40 million mark between 1968 and 1989, the amount invested per annum in Europe by Korean firms passed the US$100 million mark in 1992, and had risen to more than US$500 million within three years. Since 1986, the annual rate of growth for Korean investment in Europe has outpaced that of Korean total FDI on five occasions.

By the end of 1996, cumulative Korean investment in Europe amounted to US$2.1 billion, or 15.3 per cent of total Korean FDI in value terms. This share compared with 44.0 per cent for South-east Asia and 31.5 per cent for North America (See Table 7.2). Up to 1991, with the exception of three years (1979, 1983 and 1985), investment in Europe accounted for less than 10 per cent of global Korean FDI each year. Between 1992 and 1995, however, the annual share taken by investment in Europe rose steadily from 13.1 per cent to 21.7 per cent, before falling back to 15.8 per cent in 1996.

One notable fact concerning Korean FDI in Europe is that around 90 per cent of Korean FDI (US$1.9 billion) in the region took place between 1992 and 1996, with more than 70.5 per cent (US$1.5 billion) invested between 1994 and 1996 (Bank of Korea 1997: 26–27). This reflects the greater awareness of the importance of the region, as discussed in Chapter 6.

Investment by country

As of the end of 1996, Korean corporations had invested US$2.1 billion in 384 projects throughout Europe on a net cumulative basis. Almost two-thirds of the total (US$1.4 billion) was invested in 243 projects in western European nations: the EU member countries plus Austria, Finland, Iceland, and Malta. The remainder (US$711.7 million or 33.8 per cent of the total) went to 141 projects in central and eastern European countries (Bank of Korea 1997: 52–57). These countries are Bulgaria, the Czech Republic, Hungary, Khazakstan, Poland, Romania, the Russian Federation, Slovakia, Slovenia, Tadzikhistan, the Ukraine and Uzbekistan. After the opening up

Table 7.1 Korean Foreign Direct Investment in Europe and the World: Annual Growth Rates, 1968–1996

Year	*FDI -Europe Value (A) US$mm*	*YOY growth rate %*	*FDI - Global Value (B) US$mm*	*YOY growth rate %*	*(A) as % of (B)*
1968–1977	1.8		66.9		2.7
1978	0.9		31.4		2.9
1979	1.8	100.0	16.4	(47.8)	11.0
1980	0.4	(77.8)	12.3	(25.0)	3.3
1981	1.4	250.0	21.9	78.0	6.4
1982	1.4	0.0	97.6	345.7	1.4
1983	12.6	800.0	102.6	5.1	12.3
1984	0.9	(92.9)	48.2	(53.0)	1.9
1985	38.6	4,188.9	63.8	32.4	60.5
1986	5.5	(85.8)	158.3	148.1	3.5
1987	(42.2)	N/M	320.1	102.2	N/M
1988	15.2	N/M	156.2	(51.2)	9.7
1989	18.9	24.3	392.4	151.2	4.8
1990	90.4	378.3	812.7	107.1	11.1
1991	88.7	(1.9)	1,026.9	26.4	8.6
1992	143.4	61.7	1,097.7	6.9	13.1
1993	180.7	26.0	1,016.5	(7.4)	17.8
1994	388.6	115.1	2,028.8	99.6	19.2
1995	597.9	53.9	2,756.3	35.9	21.7
1996	557.3	(6.8)	3,530.1	28.1	15.8
Total	2,104.2		13,757.0		

Figures in parentheses indicate negative growth
N/M *Not meaningful*
YOY *Year-on-year*
Source: Bank of Korea 1997: 26–27.

Table 7.2 Korean Foreign Direct Investment: By Region, December 1996

	Net Invested US$mm	*% of Total Net Invested*
South-east Asia	6,055.4	44.0
North America	4,327.0	31.5
Europe	2,104.2	15.3
Latin America	556.0	4.0
Oceania	312.1	2.3
Africa	273.8	2.0
Middle East	128.3	0.9
TOTAL	13,757.0	100.0

Source: Bank of Korea 1997: 26–27.

of central and eastern Europe and the establishment of diplomatic relations between Korea and these nations in the late 1980s and early 1990s, there was a marked increase in Korean FDI in the region. All Korean investments in the Czech Republic, Poland, Romania, Slovakia, and Slovenia have been made since 1992, and more than 90 per cent of investment in the Russian Federation and Hungary as of the end of 1996 had taken place in the previous five years (Bank of Korea 1997: 32–34).

Despite their short history as hosts to Korean FDI, by the end of 1996 central and eastern European nations ranked high in terms of the amount invested by Korean firms. Table 7.3 shows that, in December 1996,

Table 7.3 Korean Investment in Europe: By Country, December 1996

Country	*Net Invested US$mm*	*% of Total Net Invested*	*No. of Projects*	*% of Total Projects*
United Kingdom	371.5	17.7	62	16.1
Germany	370.3	17.6	75	19.5
Netherlands	236.9	11.3	24	6.3
Uzbekistan	180.0	8.6	13	3.4
France	175.9	8.4	24	6.3
Poland	163.7	7.8	11	2.9
Romania	157.8	7.5	3	0.8
Russian Federation	107.6	5.1	79	20.6
Portugal	49.3	2.3	4	1.0
Spain	47.0	2.2	8	2.1
Hungary	40.9	1.9	8	2.1
Turkey	40.3	1.9	8	2.1
Italy	35.3	1.7	15	3.9
Tadzhikistan	29.0	1.4	1	0.3
Eire	24.9	1.2	4	1.0
Belgium	22.4	1.1	5	1.3
Slovakia	10.9	0.5	2	0.5
Austria	10.0	0.5	4	1.0
Czech Republic	10.0	0.5	4	1.0
Kazakhstan	7.8	0.4	15	3.9
Ukraine	4.1	0.2	4	1.0
Switzerland	3.4	0.2	5	1.3
Sweden	3.0	0.1	1	0.3
Malta	2.1	0.1	2	0.5
Norway	0.2	Neg.	1	0.3
Slovenia	0.1	Neg.	1	0.3
Iceland	0.04	Neg.	1	0.3
Bulgaria	0.0	0.0	0	0.0
Denmark	0.0	0.0	0	0.0
Finland	0.0	0.0	0	0.0
Total	2,104.2	100.0	384	100.0

Source: Bank of Korea 1997: 32–34.

Uzbekistan ranked fourth behind the United Kingdom, Germany and the Netherlands in terms of the value of Korean inward investment. Out of the top ten host nations at the end of that year, four were in the central and eastern European region, and Uzbekistan, Poland, Romania and the Russian Federation were among only eight out of 27 European nations to have attracted more than US$100 million in Korean inward investment. The Russian Federation accounted for the largest share of investment projects (20.6 per cent) anywhere in Europe, and Kazakhstan accounted for almost 4 per cent of Korean investment in terms of volume.

Investment by sector

As of the end of 1996, the manufacturing sector accounted for more than half the Korean capital invested in Europe (US$1.1 billion), followed by trade with a 29.5 per cent share (US$619.7 million), and mining coming a distant third (US$89.4 million for a 4.2 per cent share). Together, trade and manufacturing accounted for 83.1 per cent of Korean investment in value terms and 81.0 per cent in volume terms. Trade accounted for 170 of the 384 projects undertaken (a share of 44.3 per cent); second and third places were taken by manufacturing with 141 (36.7 per cent) and fisheries with seven (1.8 per cent) (See Table 7.4).

Manufacturing investment by country

An analysis of Korean manufacturing investment in Europe (Table 7.5) reveals a strong preference for investment in central and eastern Europe, with Uzbekistan, Poland and Romania ranked as the three most popular host nations, together accounting for more than 40 per cent of Korean investment in Europe (US$490.5 million). Six of the top ten ranked host

Table 7.4 Korean Investment in Europe: By Sector, December 1996

Sector	*Net Invested US$mm*	*% of Total Net Invested*	*No. of Cases*	*% of Total Cases*
Manufacturing	1,128.8	53.6	141	36.7
Trade	619.7	29.5	170	44.3
Mining	89.4	4.2	4	1.0
Construction	12.3	0.6	5	1.3
Transportation & Storage	2.9	0.1	5	1.3
Fisheries	2.7	0.1	7	1.8
Forestry	0.5	Neg.	2	0.5
Other	247.8	11.9	50	13.1
Total	2,104.2	100.0	384	100.0

Source: Bank of Korea 1997: 56–57.

Table 7.5 Korean Manufacturing Investment: By Country, December 1996

Country	*Net Invested US$mm*	*% of Total*	*No. of Cases*	*% of Total*
Uzbekistan	179.2	15.9	9	6.4
Romania	157.7	14.0	2	1.4
Poland	153.6	13.6	6	4.3
U.K.	123.6	10.9	31	22.0
France	105.6	9.4	12	8.5
Germany	88.4	7.8	12	8.5
Portugal	49.3	4.4	4	2.8
Russian Fed.	47.2	4.2	24	17.0
Hungary	36.5	3.2	4	2.8
Tadzhikistan	29.0	2.6	1	0.7
Turkey	25.7	2.3	5	3.5
Ireland	23.5	2.1	3	2.1
Spain	22.2	2.0	4	2.8
Netherlands	20.8	1.8	5	3.5
Belgium	16.9	1.5	2	1.4
Italy	15.0	1.3	3	2.1
Slovakia	10.8	1.0	1	0.7
Czech Republic	8.6	0.8	2	1.4
Austria	5.7	0.5	1	0.4
Kazakhstan	4.9	0.4	5	3.5
Switzerland	2.4	0.2	2	1.4
Malta	2.1	0.2	2	1.4
Slovenia	0.1	Neg.	1	0.7
Total	1,128.8	100.0	141	100.0

Source: Bank of Korea 1997: 52–57.

nations in 1996 were located in central and eastern Europe, accounting for US$603.2 million or 53.5 per cent of the total. Furthermore, of the five countries (out of 23) that attracted more than US$100 million in manufacturing investment from Korea, three were in central and eastern Europe. The United Kingdom, France and Germany took fourth, fifth and sixth places respectively and together accounted for 28.1 per cent of the total invested (US$317.6 million).

Manufacturing investment by sector

As can be seen in Table 7.6, the top five manufacturing sectors in Europe in terms of both value and volume as of the end of 1996 were fabricated metals (71.4 per cent of value and 45.4 per cent of volume), textiles and clothing (8.6 per cent and 9.2 per cent), petroleum and chemicals (6.1 per cent and 8.3 per cent), machinery and equipment (3.6 per cent and 5.7 per cent), and

Table 7.6 Korean Manufacturing Investment in Europe: By Sector, December 1996

Sector	*Net Invested US$mm*	*% of Total*	*Projects*	*% of Total*
Fabricated Metals	806.1	71.4	64	45.4
Textiles & Clothing	97.5	8.6	13	9.2
Petroleum & Chemicals	68.6	6.1	12	8.3
Machinery & Equipment	41.1	3.6	8	5.7
Primary Metals	28.8	2.6	7	5.0
Food & Beverages	24.6	2.2	8	5.7
Wood & Furniture	16.2	1.4	2	1.4
Non-metallic Minerals	9.7	0.9	1	0.7
Footwear & Leather	7.8	0.7	5	3.5
Paper & Printing	1.9	0.2	2	1.4
Other	26.5	2.3	19	13.5
Total	1,128.8	100.0	141	100.0

Source: Bank of Korea 1997: 52–57.

primary metals (2.6 per cent and 5.0 per cent). These top five sectors accounted for 92.3 per cent of FDI value and 73.6 per cent of investment volume. There was a marked gap between investment in the leading sector (fabricated metals) and the second-ranked sector (textiles and clothing) of US$806 million compared with US$98 million.

Table 7.7 shows that 15 out of 23 European countries hosting FDI from South Korea at the end of 1996 saw the largest share of that investment go to the fabricated metals sector. In six countries, this sector accounted for all the inward investment from Korea. Only four countries (Belgium, the Czech Republic, Slovenia and Tadzhikistan) saw no investment in fabricated metals. The textiles and clothing sector was the target for investment in eight countries, and accounted for all of Korean investment in Tadzhikistan. The sector also accounted for 35.3 per cent of investment in Uzbekistan and 13.7 per cent of FDI in Kazakhstan, but for only small shares (less than 4 per cent) of investment in Italy, France, the United Kingdom, Germany, and the Russian Federation.

Investment in the petroleum and chemicals sector accounted for less than 5 per cent of Korean FDI in five countries (Kazakhstan, France, the Russian Federation, the United Kingdom, and Uzbekistan) and was non-existent in thirteen others. This sector accounted for almost all investment in Eire and the Czech Republic, around 25 per cent in Belgium and Hungary, and more than 11 per cent in Germany. There was no investment in the machinery and equipment sector in 17 out of the 23 countries, and only a small amount was invested in central and eastern Europe (US$100,000 in Slovenia, which nevertheless accounted for all Korean investment in the country). The sector accounted for three quarters of investment in Belgium, more than 60 per

Table 7.7 Top Five FDI Sectors in Europe: Percentage of Investment From Korea in Each Sector, December 1996

	Fabricated Metals %	*Textiles & Clothing* %	*Petroleum & Chemicals* %	*Machinery & Equipment* %	*Primary Metals* %
Austria	100.0				
Poland	100.0				
Portugal	100.0				
Romania	100.0				
Slovakia	100.0				
Switzerland	100.0				
Italy	98.5	1.5			
Turkey	95.7				3.3
France	91.9	3.7	4.0		
Khazakstan	80.8	13.7	2.0		3.0
Germany	72.7	0.1	11.3	2.4	
Uzbekistan	62.2	35.3	2.5		
Hungary	59.6		25.3		
Netherlands	59.1			14.5	0.2
UK	55.4	0.1	3.5	17.7	0.8
Malta	39.2			60.8	
Russian Fed.	11.2	0.3	1.9		31.5
Spain	5.2				53.9
Eire	4.3		95.6		
Tadzhik.	0.0	100.0			
Czech Rep.	0.0		99.5		
Belgium	0.0			75.1	
Slovenia	0.0		24.9	100.0	

Source: Bank of Korea 1997: 90–95.

cent in Malta, between 14 per cent and 18 per cent in the Netherlands and the United Kingdom, and over 2 per cent in Germany. Investment in primary metals was concentrated in six countries, accounting for more than half of Korean investment in Spain, 31.5 per cent in the Russian Federation, around 3 per cent in Turkey and Kazakhstan, and less than 1 per cent in the United Kingdom and the Netherlands.

Table 7.8 shows that the top seven host nations for South Korean investment in the trading sector were located in western Europe, accounting for more than 95 per cent of the amount invested and almost 70 per cent of investment projects undertaken. Between them, Germany and the United Kingdom accounted for three quarters of the amount invested by Korean firms in trade-related projects. In total, western European nations hosted investment valued at more than US$600 million, or 97 per cent of the total.

Table 7.8 Korean Trading Investment by Country, December 1996

Country	*Net invested US$mm*	*% of Total*	*No. of Projects*	*% of Total*
Germany	276.5	44.6	52	30.6
United Kingdom	191.4	30.9	24	14.1
France	66.0	10.7	10	5.9
Spain	24.8	4.0	4	2.4
Italy	20.0	3.2	10	5.9
Netherlands	7.9	1.3	14	8.2
Belgium	5.5	0.9	3	1.8
Russian Fed.	5.3	0.9	26	15.3
Hungary	4.4	0.7	4	2.4
Austria	4.1	0.7	1	0.6
Poland	3.2	0.5	4	2.4
Sweden	3.0	0.5	1	0.6
Ukraine	2.6	0.4	3	1.8
Kazakhstan	1.6	0.3	6	3.6
Czech Rep.	1.3	0.2	2	1.2
Switzerland	1.0	0.2	2	1.2
Uzbekistan	0.7	0.1	1	0.6
Turkey	0.3	0.1	1	0.6
Slovakia	0.1	Neg.	1	0.6
Iceland	0.04	Neg.	1	0.6
Total	619.7	100.0	170	100.0

Source: Bank of Korea 1997: 52–57.

Table 7.9 Korean Investment in Western, Central and Eastern Europe: By Sector, December 1996

	W. Europe		*C&E. Europe*	
	Net Invested US$mm	*Cases*	*Net Invested US$mm*	*Cases*
Mining	64.3	3	25.1	1
% of Total	71.9	75.0	28.1	25.0
Forestry	0.0	0	0.5	2
% of Total	0.0	0.0	100.0	100.0
Fishing	0.2	1	2.5	6
% of Total	7.4	14.3	92.6	85.7
Manufacturing	501.2	86	627.6	55
% of Total	44.4	61.0	55.6	39.0
Construction	5.0	2	7.3	3
% of Total	40.7	40.0	59.3	60.0
Transportation & Storage	2.9	5	0.0	0
% of Total	100.0	100.0	0.0	0.0
Trade	600.5	123	19.2	47
% of Total	96.9	72.4	3.1	27.6
Other	218.4	23	29.5	27
% of Total	88.1	46.0	11.9	54.0

Source: Bank of Korea 1997: 52–57.

As can be seen in Table 7.9, in addition to taking the lion's share of Korean investment in trade (96.9 per cent), western European nations have also received more than 70 per cent of investment in the mining sector and all South Korean investment in transportation and storage. The central and eastern European nations have hosted all the investment in forestry, and the majority of Korean FDI in fisheries (92.6 per cent), construction (59.3 per cent), and manufacturing (55.6 per cent). Within the manufacturing sector, western European nations accounted for the largest share of Korean investment, in terms of value, in paper and printing, non-ferrous metals, machinery and equipment, footwear and leather, food and beverages, petroleum and chemicals (See Table 7.10).

The statistics reflect a trend towards building large-scale facilities requiring high technology in advanced countries, with the Koreans investing, for example, in machinery and equipment, petroleum and chemicals in the United Kingdom, Germany, France, and Belgium, while

Table 7.10 Korean Manufacturing Investment in Western, Central and Eastern Europe: By Sector, December 1996

	W. Europe		*C&E. Europe*	
	Net Invested US$mm	*Cases*	*Net Invested US$mm*	*Cases*
Food & Beverages	15.2	3	9.4	5
% of Total	61.8	37.5	38.2	62.5
Textiles & Clothing	4.3	6	93.2	7
% of Total	4.4	46.2	95.6	53.8
Footwear & Leather	7.5	2	0.3	3
% of Total	96.3	40.0	3.7	60.0
Wood & Furniture	0.0	0	16.2	2
% of Total	0.0	0.0	100.0	100.0
Paper & Printing	1.9	2	0.0	0
% of Total	100.0	100.0	0.0	0.0
Petroleum & Chemicals	45.3	6	23.2	6
% of Total	66.1	50.0	33.9	50.0
Non-Ferrous Metals	9.7	1	0.0	0
% of Total	100.0	100.0	0.0	0.0
Primary Metals	13.8	4	15.0	3
% of Total	47.9	57.1	52.1	42.9
Fabricated Metals	341.8	44	464.3	20
% of Total	42.4	68.8	57.6	31.2
Machinery & Equipment	41.0	7	0.1	1
% of Total	99.8	87.5	0.2	12.5
Other	20.9	11	5.6	8
% of Total	78.7	57.9	21.3	42.1

Source: Bank of Korea 1997: 90–95.

concentrating labour-intensive activities in low wage-cost nations. Examples of the latter were leather and footwear investments in Spain, and fabricated metals projects in Portugal (Bank of Korea 1997: 90–95). The Russian Federation hosted all the Korean investment in the wood and furniture sector and the forestry sector, while central and eastern European nations received the majority share of investment in the textiles and clothing sector (95.6 per cent), fabricated metals (57.6 per cent), and primary metals (52.1 per cent). The patterns of investment largely reflected the Koreans' search for low wages for labour-intensive industrial production in the textiles and fabricated metals sectors (Bank of Korea 1997: 90–95).

It is interesting to note that, although only small amounts of investment had been made in Kazakhstan in absolute terms (0.4 per cent of total manufacturing investment in Europe), Korean companies were active there in five out of the ten sectors, namely food and beverages, textiles and clothing, petroleum and chemicals, primary metals, and fabricated metals. Of the total, more than 80 per cent of investment in Kazakhstan was made in fabricated metals production facilities. Only the Russian Federation had a more varied investment portfolio, with a Korean presence in seven out of ten sectors (excluding paper and printing, non-ferrous metals, and machinery and equipment). Almost two-thirds of Korean investment in the Russian Federation was in resource-seeking projects (Bank of Korea 1997: 90–95).

Investment by project size and equity participation ratio

Table 7.11 reveals a difference in the scale of investment projects undertaken in Europe and North America and those implemented in South-east Asia. At the end of 1996, projects valued at more than US$5 million accounted for 89.3 per cent of the total in Europe and 84.2 per cent of the total in North America, but only 53.4 per cent in South-East Asia. Table 7.12 shows the

Table 7.11 Average Size of Korean Investments in Europe, North America and South-east Asia, December 1996

Size	*Europe* %	*North America* %	*SE Asia* %
<US$1 million	1.3	3.8	17.6
US$1–2 million	2.0	3.4	12.3
US$2–5 million	7.4	8.6	16.7
US$5–10 million	11.0	8.2	13.8
>US$10 million	78.3	76.0	39.6
Total	100.0	100.0	100.0

Source: Bank of Korea 1997: 78.

Table 7.12 Equity Participation Ratio in Europe, North America and South-east Asia, December 1996

EP Ratio	*Europe* %	*North America* %	*SE Asia* %
<20%	1.7	3.0	1.2
21–49%	14.1	11.2	10.5
50%	14.7	0.3	4.9
51–99%	37.5	34.9	41.0
100%	32.0	50.6	42.4
Total	100.0	100.0	100.0

Source: Bank of Korea 1997: 113.

preference for majority share holding which is characteristic of Korean FDI worldwide. In more than 80 per cent of cases in all three regions, the Korean parent company held an ownership stake of 50 per cent or above. The share was 84.2 per cent in Europe, 85.8 per cent in North America, and 88.3 per cent in South-east Asia.

Korean investment in European real estate has been minimal: at the end of 1996, a total of US$6.2 million had been invested in four projects. In 1991, the investment total stood at US$6.5 million in five projects but 1993 saw the withdrawal of US$0.3 million and the disposal of one asset, and there were no acquisitions between 1994 and 1996 (Bank of Korea 1997: 126).

Motivations for investment

Korean investment in Europe presents a somewhat complex character, combining as it does investment to secure and expand markets, investment to take advantage of low-cost labour, and investment to acquire advanced technology and managerial expertise.

Western Europe

One of the most important motivations for Korean firms investing in western Europe has been the need to defend this important export market. As Europe moved towards the establishment of a single market, an additional motivation emerged, namely that of securing a production base within the bloc and thus receiving the benefits of integration. The establishment of a manufacturing base in Europe could also create opportunities for sourcing key supplies from local SMEs; high domestic interest rates and restricted sources of credit served as a further motivation to invest in countries with access to low-cost finance (Randerson and Dent 1996: 50, 69). Finally, in the era of globalization, Korean investors have been motivated by the desire to establish a global corporate structure, and acquire advanced technology and management skills.

These changes in motivation were reflected in shifts in the method of entry used; in the 1970s, the Koreans set up sales subsidiaries in major European cities but, as Europe moved towards integration, it became clear that the strategy of advancing into the EU market through exports would have its limitations. Therefore, in order to secure the market, avoid trade restrictions and respond quickly to the demands and tastes of local consumers, it was essential to establish a local production base in Europe (Wang 1997:180–181).

Central and Eastern Europe

A report published by UNCTAD in 1997 identified the search for low-cost labour as a key motivation for Korean investment in central and eastern Europe. Another important factor was gaining access to the western European market by a 'back door' route through central and eastern Europe. Excellent examples of this strategy have been Daewoo's acquisition of automobile plants in Romania, Poland and the Czech Republic, and the purchase of production facilities being privatized by eastern European governments. (It was via this route that Daewoo acquired a television manufacturer in East Berlin.)

A third motivation has been current or anticipated trade friction and restrictions; many countries in central and eastern Europe enjoy preferential access to the EU market and Korean firms could avoid import restrictions imposed in the EU by manufacturing in and exporting from central and eastern European countries. A final factor was the potential demand in the markets themselves, as economic reforms were carried out and disposable incomes rose (UNCTAD 1997: xv, 30–32, 37).

Jeong (1998) noted the strong correlation between Korean trade with and investment in central and eastern European countries; just as electronics appliances, automobiles and textiles together accounted for between 85 per cent and 95 per cent of Korean exports, investment in the region has been led by the electronics and automobiles (fabricated metals) sector (Jeong 1998: 29–30). A final point worth noting is the presence of Korean ethnic minorities in some of the countries of central Europe, which may serve as a secondary incentive in choosing an investment location.

KOREAN CONSUMER ELECTRONICS INVESTMENT

Writing in 1997, Ungson *et al* noted the importance of globalization as an element of the strategies developed by Korean corporations in response to changes in the domestic market, the accelerating liberalization and deregulation of global markets, and the need to acquire advanced technology and expertise. The process of globalization begins with domestic market orientation (domestic sales), moves on to international market orientation (boosting production capacity at home and opening sales offices overseas), and then to the establishment of a company overseas, which employs local staff and manufactures goods using locally-sourced parts and components. The final stage is the establishment of a company with a truly global organizational structure and flexible, efficient operations (Ungson *et al* 1997: 110–114).

For Korea's consumer electronics companies the process of globalization began in 1962, when Goldstar exported its first shipment of radios to Hong Kong. Goldstar moved swiftly to establish a presence in the American and European export markets in the 1960s, with Samsung and Daewoo following over the next decade. Local sales subsidiaries were set up in the United States in the late 1970s and in Europe during the early 1980s. This activity reflected that fact that, as has been noted in Chapter 4, the focus of export activity within the consumer electronics industry in the 1970s and 1980s was North America and the countries of western Europe. It was only at a later stage that Korean consumer electronics firms established sales subsidiaries in Asia and eastern Europe. Investment in manufacturing facilities – the next stage of globalization – began in 1982, when Goldstar established a manufacturing base in the United States. Although Samsung began production in Europe shortly afterwards, once again the entry into Asia took place at a much later stage.

The tables in this section were compiled using data published by the Electronics Industries Association of Korea (EIAK) in their annual directory of Korean electronics firms operating subsidiary companies overseas. Although the data may be incomplete (only data pertaining to EIAK member companies is published, and some members provide only the minimum information requested), this appears to be the most comprehensive picture of FDI by Korean electronics firms published in Korea. It should be noted that volume and value of investment are given on an approval basis, which is always greater than the amounts actually invested. Nevertheless the statistics give a good picture of the trends in Korean electronics investment overseas. It should also be noted that the data given by the EIAK relate to the investment position at the end of June 1997.

Korean electronics investment – Worldwide

Between 1968 and June 1997, Korean electronics firms received approval for a total of 1,204 overseas investment projects, valued at US$5.9 billion. (On an invested basis, the amount was US$3.5 billion in 1,014 projects.) More than half the amount approved (US$3.4 billion or 57.6 per cent) was in manufacturing, with US$1.7 billion (28.8 per cent) to be invested in sales, and US$125 million (2.1 per cent) in R&D (EIAK 1997: 259). Table 7.13

Table 7.13 Korean Global Electronics FDI Projects: By Sector, June 1997

	Total	*Consumer Electronics*	*Industrial Electronics*	*Parts & Components*	*Other*
Manufacturing	784	170	168	439	7
Sales	306	135	66	92	13
R&D	33	3	1	4	28
Other	81	0	5	2	71
Total	1,204	308	240	537	119

Source: EIAK 1997f: 259.

Table 7.14 FDI Projects by Korean Electronics Companies, June 1997

	1968–1989	*1990*	*1991*	*1992*	*1993*	*1994*	*1995*	*1996*	*Total*
Asia	24	28	39	45	61	140	128	126	591
N. America	12	6	5	7	8	16	21	25	100
C/L. America	1	2	0	0	1	1	5	9	19
Europe	12	4	3	7	8	5	14	15	68
Oceania	1	0	0	0	0	1	0	0	2
Africa	1	1	0	0	0	1	0	1	4
Manuf'g Total	**51**	**41**	**47**	**59**	**78**	**164**	**168**	**176**	**784**
Asia	28	7	6	6	8	22	20	19	116
N. America	28	6	5	3	9	10	17	8	86
C/L. America	4	0	0	1	2	5	2	4	18
Europe	15	6	6	10	8	5	11	10	71
Oceania	4	1	0	0	0	1	2	1	9
Africa	0	0	0	0	1	0	2	3	6
Sales Total	**79**	**20**	**17**	**20**	**28**	**43**	**54**	**45**	**306**
Asia	0	4	2	2	2	3	0	0	13
N. America	1	4	1	2	1	5	1	1	16
C/L. America	1	0	0	0	0	0	0	0	1
Europe	0	1	1	0	0	1	0	0	3
R&D Total	**2**	**9**	**4**	**4**	**3**	**9**	**1**	**1**	**33**
Asia	0	0	1	0	3	6	7	12	29
N. America	2	1	1	1	2	3	7	20	37
C/L. America	0	0	0	0	0	0	1	1	2
Europe	0	0	0	0	0	(1)	6	4	11
Other	**2**	**1**	**2**	**1**	**5**	**8**	**23**	**37**	**81**
Total	134	71	70	84	114	226	246	259	1,204

Source: EIAK 1997f: 260.

shows that, in terms of volume, the parts and components sector accounted for the largest share of projects approved (537 cases, 44.6 per cent). Consumer electronics accounted for 25.6 per cent (308 cases), and industrial electronics a further 19.9 per cent (240 cases).

As can be seen in Table 7.14, another important point highlighted by the statistics published by the EIAK is the rapid increase in FDI by Korean electronics companies in the early 1990s. Among the 1,204 projects authorized by the Korean authorities, more than 60 per cent (731 cases) were approved between 1994 and 1996. This increase in investment in the mid-1990s was particularly evident in the manufacturing sector, with 65 per cent of projects being approved between 1994 and 1996. In Central and Latin America, almost 80 per cent of projects were approved in this three-year period (15 projects out of 19); the percentages for Asia, North America, and Europe were 66.7 per cent (394 out of 591), 62.0 per cent (62 out of 100), and 50 per cent (34 out of 68).

The rapid increase in FDI activity in the mid-1990s reflects the efforts made by Korean electronics firms to respond to changes in the domestic and global environment. Rising labour costs at home prompted firms to relocate production to regions which still offered the advantage of low-cost labour, while the desire to overcome trade restrictions and gain access to technology and expertise as part of a wider globalization strategy encouraged investment in the industrialized markets of the United States and western Europe.

Out of the 1,204 projects approved, 30 per cent were to be located in the People's Republic of China (PRC); the leading position assumed by the PRC was particularly remarkable in that diplomatic relations between China and South Korea had only been established in 1992. North America was to be host to 227 projects, or almost 19 per cent of the total. Of the top five host countries, four were in Asia, and only two western European countries were listed in the top ten (See Table 7.15). Korean electronics companies had begun to invest heavily in low-cost production locations in Asia; the PRC's ranking as the most favoured host country reflects its abundance of low-cost manpower, the medium- to long-term domestic market demand for lower-end electronics products, and also (perhaps to a lesser degree) the presence of a sizeable ethnic Korean community in the north-eastern region of the country.

Korean consumer electronics investment in Europe

As has been noted earlier, by the late 1990s, the European Union had emerged as the largest host to foreign direct investment in the world, accounting for almost 40 per cent of global investment flows. Nevertheless, a report published by the United Nations Conference on Trade and Development (UNCTAD) in 1997 noted the low levels of Asian investment in Europe, which it attributed to a number of factors. These included the traditional orientation of Asian firms towards Japan and the United States, the need for a strong competitive edge when investing in industrialized markets, the cost of collecting vital market information, and the burden of

Table 7.15 FDI by Korean Electronics Companies: By Country, June 1997

Host Country	*Cases*	*% of Total*
People's Republic of China	363	30.1
North America	227	18.9
Philippines	98	8.1
Hong Kong	68	5.6
Japan	49	4.1
Malaysia	36	3.0
Indonesia	31	2.6
United Kingdom	31	2.6
Germany	29	2.4
Russia	21	1.7
Singapore	15	1.2
Mexico	14	1.2
Australia	11	0.9
France	11	0.9
Canada	10	0.8
Others	190	15.9
Total	1,204	100.0

Source: EIAK 1997f: 261.

transaction costs in Europe (including internal transportation, communication, requirement to comply with technical standards and so forth).

The report observed that most European governments had taken steps to deregulate and liberalize inward investment, improve the investment environment, and provide incentives that would attract foreign capital and create jobs. However, this action to boost inward investment into Europe had not, generally speaking, been matched by efforts to promote investment in Europe on the part of Asian governments. UNCTAD observed that, although many governments (including that of Korea) had eased restrictions on outward FDI, the support offered to potential investors was seldom specifically targeted in favour of Europe (UNCTAD 1997: 47–62).

The UNCTAD report also highlighted the need for Korean firms to invest in Europe, arguing that a local presence was essential in order for foreign manufacturers to keep up with trends in consumer taste and changes in buying patterns, and thus develop products suited specifically to the market. A local presence could also allow access to advanced technology through joint ventures and technical alliances with European firms, the recruitment of specialist staff, the absorption of knowledge through contacts with local suppliers and consultants, or the acquisition of small, technology-based companies. Close contact with European firms could also enhance understanding of their corporate culture and managerial methods which would help Korean firms achieve their goals of globalization and internationalization.

In terms of production costs, the report pointed out that, although average wage levels were higher in western Europe than in Korea, there were low wage areas within the EU and in eastern Europe, and wage differentials within the EU could be expected to fall in line with increasing labour market flexibility, persisting unemployment and the shift to low-wage service employment. The combination of good infrastructure and access to networks of high quality suppliers and skilled workers made Europe a excellent base for production which was not labour cost sensitive. Furthermore, inward investment that boosted the local economy and created jobs could help to ease protectionist sentiments toward and trade friction with Korea (UNCTAD 1997: 64–68).

In a study of the competitiveness of the Korean consumer electronics industry published in 1990, Sŏnu *et al* identified a number of factors contributing to the strategic importance of the European market for Korean consumer electronics firms, which sheds light on the Korean perspective as regards the importance of the region. Firstly, Europe was the second largest market for consumer electronics products in the world: in 1988, demand for consumer electronics products in the EC's 14 member states was US$15.5 billion, accounting for 29.5 per cent of the global market (valued at US$52.8 billion).

Secondly, the European market offered Korean firms the opportunity of reducing their traditional export dependence on the United States. Finally, by investing in Europe, consumer electronics firms could create a 'strategic bridgehead' for forging closer trade and investment links with central and eastern Europe. The authors took the view that it was 'impossible' to attack the European market through exports, due to trade friction and increasing demands for localization; direct investment in the market was, therefore, essential (Sŏnu *et al* 1990: 340–342).

Motivations for investment

In 1994, Chi Nam-ung published a study of the internationalization process and strategy of the Korean consumer electronics industry, in which he considered the motivations for investment in the latter half of the 1980s and early 1990s. From interviews with businessmen and industry experts carried out in 1993, Chi concluded that early FDI projects were undertaken from defensive motivations such as avoiding trade restrictions, defending existing export markets, securing a base for reverse exports and so forth. (These findings were very much in line with Wŏn's study of investment by Korean consumer electronics companies in the United States in the early 1980s.)

As the domestic and global environments began to worsen from the latter half of the 1980s, with increasing protectionism, the emergence of regionalism, rising wages at home and the erosion of price competitiveness, Korean electronics firms maintained or expanded their investments in industrialized countries, despite the loss of their competitive advantage. At the same time, however, they began investing in developing countries and the former Eastern bloc in search of low-cost labour, new product demand and bases for exports to third countries.

According to Chi, in this 'new era of FDI', the key motivations were, in order of importance, to gain access to low-cost local labour, to secure a base for exports into third countries, to open up new markets, to promote the growth of the firm, to respond to a competitor's move overseas, to protect existing markets, to exploit a firm-specific advantage, to gain access to local raw materials, to transfer idle equipment overseas, and to avoid import restrictions. The top two motivations reflect the fact that the majority of Korean manufacturing investment was now going to developing countries (Chi 1994: 86–100). It is interesting to note that oligopolistic reaction appeared in the list of motivations; this is a point to which we will return in the next chapter.

In addition to the general motivations for investment by consumer electronics companies identified by Chi, two recent studies have highlighted motivations specific to investment by these firms in the European Union. According to Byun (1995), out of all the industrial sectors facing anti-dumping measures in the 1980s and 1990s, the electronics sector was hit the hardest, with 22 cases against South Korea, 24 against Japan, and 31 against China. There were EU-wide restrictions on 12 electronics products, including anti-dumping duties on compact disc players, car stereos, colour televisions with screens smaller than 16 inches, and video and audio tapes. Investigations were underway with regard to the possible dumping of larger colour televisions; price undertakings had been agreed for video cassette recorders, and quotas had been imposed on microwave ovens and colour picture tubes (Byun 1995: 135–140, 165–186). The imposition of import restrictions on major export items posed a threat to the firms' growth and, arguably, their very survival, making a local presence in the market essential. A study published by Shin in 1998 noted the important role played by EU policy relating to the consumer electronics industry which, he claimed, had 'forced' Korean and Japanese firms to locate production plants within the EU. According to Shin, in addition to pressure from local firms, the European Commission had been lobbied by Japanese firms seeking protection from the export offensive by the East Asian NIEs in the 1980s (Shin Sang-hyup 1998: 99–102).

CONCLUSIONS

The analysis of the development of diplomatic and economic relations between Korea and Europe in Chapter 6 revealed that it is only since the late 1980s that strong bilateral links have been developed. In the case of Korean direct investment in Europe, it has been shown that the majority of Korean

FDI activity in the region took place in the five-year period between 1992 and 1996. During the 1970s and early 1980s, Korean investment activity was focused on South-east Asia and North America; Korean FDI in Europe was limited to the establishment of a network of sales facilities in western Europe, by means of which Korean exports to the region could be promoted and expanded. It was not until the latter half of the 1980s that Korean manufacturing investment in Europe began in earnest.

The success of Korean exports to western Europe can be judged by the speed and severity of the import restrictions placed upon them. The governments of major western European countries moved swiftly to stem the influx of low-priced Korean products and exporters found a key market under threat. Concerns over the further strengthening of these restrictions, and fears that firms failing to establish a presence in the EU would find themselves locked out of 'Fortress Europe' prompted Korean companies to develop an investment strategy for Europe.

Additionally, in the late 1980s and early 1990s, investors were seeking opportunities to acquire advanced technology through collaborative agreements, joint ventures or M&A, and to incorporate Europe into their growing global networks of production and distribution. Since 1992, Korean firms have advanced into the countries of central and eastern Europe, attracted by low production costs, local market potential, and access to the countries of the European Union.

Korean investment in Europe has been dominated by companies in the manufacturing and trade sectors, which together accounted for more than 80 per cent of the funds invested as of the end of 1996. By far the largest share of manufacturing investment has gone to the fabricated metals sectors, which, although including automobile sector investments made in the mid-1990s, has been dominated by the consumer electronics firms. The fabricated metals sector accounted for between half and all of the investment made by Korean firms in 15 out of the 23 countries included in the official FDI data. This reflects the electronics industry's leading position among Korea's export sectors and the efforts made by manufacturers to defend and expand this key market.

Another interesting feature of Korean investment in Europe is the extremely high concentration of trading investment in the countries of western Europe: eight out of the top ten host countries for trading investment were industrialized countries. This reflects, in part, the longer trading history between Korea and western Europe but, more importantly, the size and actual (rather than potential) consumer demand in these countries. The countries of central and eastern Europe have taken the lion's share of investment in manufacturing (the search for lower production costs) and forestry (the search for natural resources).

Despite the preference for central and eastern Europe for labour-intensive manufacturing, there has also been a tendency for the Koreans to focus their technology- and capital-intensive manufacturing in sectors such as machinery and equipment, petroleum and chemicals in western Europe. Finally, the marked difference in average project size between investments made in Europe and those undertaken in South-east Asia (almost 90 per cent valued at more than US$5 million as opposed to just over 53 per cent) shows a trend towards building large-scale facilities in Europe, be they technology-intensive projects in western Europe or labour-intensive facilities in central and eastern Europe.

The data presented in this section supports the view that Korean FDI in Europe has been motivated by changes in the domestic and global business environment. The periods of active investment coincide with phases in Korea's economic development in which significant changes were taking place. Korean manufacturers and exporters established a sales presence in Europe during a period of intense export promotion, which included the creation of the GTCs, who spearheaded this advance. Manufacturing investment in western Europe, which began as the effects of trade friction began to be felt, accelerated as Europe moved towards integration and further intensified as Korean companies geared up for the era of globalization and borderless competition. Strategic investment, aimed at the creation of global production and sales networks and the acquisition of advanced technology and management skills, became a feature of Korean FDI in the early 1990s as the process of industrial restructuring gathered momentum.

The final chapter of this book draws together all the factors that have served to promote FDI by Korean firms in general, and by consumer electronics companies in particular, to create a theoretical/macroeconomic framework within which this investment may be understood.

Chapter Eight

Korean Consumer Electronics Investment in Europe: A Theoretical/Macroeconomic Perspective

The purpose of this study is to construct a theoretical/macroeconomic framework within which Korean consumer electronics investment in Europe may be understood. In order to create this framework, we have examined western and Korean theories of foreign direct investment, and assessed the impact of Korea's economic development process and industrial structure on the ability of Korean consumer electronics companies to develop firm-specific monopolistic advantages. We have also analyzed the various phases of Korean FDI, linking them with the stages in Korea's economic development, and have related patterns of investment in Europe to changes in the global business environment and the bilateral relationship.

In the early chapters of the book, the theories developed by western and Korean scholars to explain the phenomenon of foreign direct investment were presented and analyzed. Particular focus was placed on the explanation of reverse (or upstream) direct investment, of which Korean investment in the industrialized countries of western Europe is a prime example.

A survey of the Korean literature relating to foreign direct investment in general, and overseas investment by Korean consumer electronics companies in particular, has revealed a consensus among scholars on the following points. Firstly, that western and Japanese theories of FDI offer only a partial explanation of reverse direct investment. Secondly, that RDI by Korean consumer electronics companies can be characterized as passive, immature, and defensive investment which has taken place in the absence of a clear firm-specific monopolistic advantage. Thirdly, that Korean reverse investment has been undertaken as a matter of survival in the face of changes in the domestic and global environment. Finally, some scholars have concluded that the oligopolistic structure of the Korean consumer electronics industry has had a major influence on RDI activity, leading to 'follow-the-leader' patterns of investment.

According to conventional theories of foreign direct investment, an essential prerequisite for firms wishing to engage in investment overseas is the possession of a firm-specific monopolistic advantage. This advantage, which may take many forms including superior technology, a famous brand name or trade mark, and access to low-cost financing, enables the investor to offset the 'cost of foreignness' inherent in operating in a foreign market, and compete with local firms or other multinational enterprises in the market.

Although Korean firms can be said to have advantages when investing in developing country markets (through the possession of technology appropriate to local market conditions, experience in operations in developing countries and so forth), it is generally agreed that no such advantage was in evidence in the case of Korean companies investing in industrialized countries in the late 1980s and early 1990s. Writing in 1987, Jun Yongwook observed that Korean consumer electronics firms investing in Europe in the 1980s were doing so in the absence of a clear firm-specific monopolistic advantage. These companies had no brand image, their technology and R&D investment levels were lower than those of their competitors in Europe, they had limited access to distribution and marketing, a weak parts and components industry, and the burden of high cost of capital. Jun characterized the investments in Europe as 'market security-seeking', aiming to bypass anticipated import barriers and thus protect key export markets (Jun 1987: 97–98).

Yun Tong-jin concurred with Jun on the lack of firm-specific monopolistic advantage, and put forward the view that such advantages as the Koreans did enjoy were comparative advantages such as the sourcing of cheap parts and components from their home market. However, those advantages could easily be eroded by changes in the domestic and global environments, including exchange rate fluctuations and rising wages (Yun 1993: 242).

From our analysis of the development and growth of the consumer electronics industry, it has become clear that the characteristics of the industry have hampered the acquisition of firm-specific advantages to a considerable extent. In the early stages of the development of the consumer electronics industry, companies were encouraged by the government to achieve price competitiveness on the basis of mass production, taking advantage of Korea's abundant pool of low-cost labour.

The emphasis on rapid growth and export promotion, coupled with the presence of foreign companies in the Korean market led to a heavy dependence on external sources for technology and expertise. The fact that foreign firms could provide not only capital and technology but also indirect access to research, design, marketing and distribution capabilities through OEM sales, robbed the consumer electronics companies of the motivation to invest time and money in the development of these key processes in the value-added chain. Thus it can be said that the foundation of the Korean consumer electronics industry was adaptation rather than innovation. This lack of focus and investment left the consumer electronics industry with a weak technology base and an underdeveloped parts and components industry. Furthermore, companies that had relied on OEM sales in the 1970s and early 1980s found themselves without a brand name or corporate image in global markets in the late 1980s and early 1990s.

A recurring theme in Korean surveys of foreign direct investment is the applicability of Knickerbocker's Oligopolistic Reaction Theory to Korean patterns of investment. As was discussed in Chapter 2, Wŏn Chong-gŭn (1986) incorporated the theory into his theoretical explanation of RDI, and other authors including Jun Yongwook (1988) and Park Gi-han (1996) have incorporated this approach into their studies of RDI.

Jun analyzed the time lags between critical points in the internationalization process (from exports to local sales through a subsidiary and local production) followed by Goldstar and Samsung. He found that, generally speaking, the gaps were very close; many were less than a year and the longest was just three years. Jun also found a 'striking similarity' in the locations of overseas subsidiaries and branch offices. The two companies also took turns in being the leader, with Goldstar taking the lead in exports in the 1970s until Samsung achieved a strong market position with new products such as the microwave oven, and advanced into new markets overseas.

The moves demonstrated that, after the successful entry into an overseas market by one firm, a rival might follow to take advantage of the leader's experience. This might, in turn, prompt the leader to expand the original investment or make another advance overseas. Jun concluded that, in addition to trading barriers erected in key markets, rivalry between firms in an oligopolistic industry is a 'very important factor' in explaining Korean FDI in industrialized countries (Jun 1988: 60–70).

Park's 1996 survey of investment in the United States, United Kingdom and Germany by LG and Samsung also revealed evidence of follow-the-leader patterns of investment in exports, local sales and local production. Park reported that the time lags between moves made by LG and Samsung were, generally speaking, 'relatively short', averaging between a few months and three years (Park Gi-han 1996: 321).

The general consensus that Korean firms invested in industrialized countries in the absence of a clear firm-specific monopolistic advantage supports our belief that companies were forced to invest in these markets by changes in the domestic and global business environment. These changes are closely related to the economic development of Korea and the growth and characteristics of the consumer electronics industry.

The remainder of this chapter will present a detailed analysis of Korean consumer electronics investment in Europe, using data for western, central and eastern Europe through to the end of June 1997. The applicability of the oligopolistic reaction theory to this area of Korean FDI will be tested against the patterns of investment by the three giant consumer electronics companies, Samsung Electronics, LG and Daewoo Electronics. Having considered Korean FDI in Europe in the context of this western theory, we will expand upon the concepts of 'structural irreversibility' and 'internalization of external factors' put forward by Jun Yongwook and Yun Tong-jin to create a theoretical/macroeconomic framework for Korean FDI. The framework will incorporate the various domestic and global factors that have driven overseas investment by Korean consumer electronics firms since the late 1980s.

KOREAN CONSUMER ELECTRONICS INVESTMENT IN EUROPE: PLAYING 'FOLLOW-THE-LEADER' OR 'LEAPFROG'?

From the early days of FDI in Europe by Korean consumer electronics companies, investment was led and dominated by the 'Big Three' consumer electronics companies, namely LG, Samsung, and Daewoo, whose investment forms the focus of this section.[1] The data, once again, has been obtained from the Electronics Industries Association of Korea and includes investments made up to the end of June 1997. An important omission from the data presented in the EIAK directory is that concerning joint ventures between Korean and European firms. Although such data as was available from other sources has been incorporated into the tables, it is possible that more joint ventures have been established by Korean firms than are shown here.

In 1980, Goldstar became the first Korean consumer electronics company to advance into Europe, setting up a sales facility in Germany. Two years later, Samsung Electronics set up the first Korean electronics manufacturing facility in Europe, taking a 55 per cent holding in a joint venture colour television factory in Portugal (Yun 1993: 283–284). Within two decades, Korean consumer electronics firms had established a network of sales and manufacturing operations throughout Europe, starting in western Europe in the 1980s and moving into central and eastern Europe in the late 1980s and early 1990s.

As of the end of June 1997, LG maintained sales outlets in Germany (established in 1980), the United Kingdom (1986 and 1996), France (1990) and Hungary (1992), and operated manufacturing facilities in Germany (1986), Turkey (1987), the United Kingdom (1988), and Italy (1990). Samsung operated a broader sales network, encompassing Germany and Portugal (1982), the United Kingdom (1984), France (1988), Spain (1990), Italy (1991), Sweden (1992), the Netherlands (1995 and 1997), Russia (1995), and Poland (1996). Samsung's manufacturing facilities were located in Portugal (1982), the United Kingdom (1987), Spain, Hungary, Italy and Turkey (1989).

While LG and Samsung focused their FDI activity in western Europe, Daewoo took the lead in establishing a sales and manufacturing presence

Table 8.1 Investment in Europe by the 'Big Three' Consumer Electronics Companies, June 1997.

	LG		*Samsung*		*Daewoo*	
	Sales	*Prod'n*	*Sales*	*Prod'n*	*Sales*	*Prod'n*
France	1990		1988		1991	1988 1992 1993
United Kingdom	1986 1996	1988	1984	1987	1993	1988
Spain			1990	1989	1992	1996
Germany	1980	1986	1982		1992	
Hungary	1992			1989	1993	
Uzbekistan						1993
Poland			1996		1993	1993
Russian Federation			1995		1993 1996	
Czech Republic					1993	
Netherlands			1995 1997		1994	1996
Kazakhstan						1994
Italy		1990**	1991	1989**	1994	
Romania					1995	
Ukraine					1996	
Portugal			1982	1982**		
Sweden			1992			
Turkey		1987**		1989**		

** Joint venture.

Sources: EIAK 1997f, Park Gi-han 1996: 321, and Jun and Simon 1992: 196.

in eastern Europe. (Although Samsung was the first Korean consumer electronics firm to establish a presence in eastern Europe with a production facility in 1989, its investment activity in the region has been limited to the addition of two sales subsidiaries in Poland and Russia.) Another notable aspect of Daewoo's investment is that, apart from its manufacturing plants in France and the United Kingdom, which were established in 1988, all the firm's investments in Europe took place in the 1990s.

As of the end of June 1997, Daewoo had set up sales subsidiaries in France (1991), Spain and Germany (1992), the United Kingdom, Hungary, Poland, Russia, the Czech Republic (1993), the Netherlands and Italy (1994), Romania (1995), and the Ukraine (1996). A second Russian sales outlet was set up in 1996. In the 1990s, Daewoo began production in Uzbekistan and Poland (1993) and Kazakhstan (1994); additional manufacturing facilities were established in France in 1992 and 1993. Production in Spain and the Netherlands began in 1996.

In his 1996 study of the sales and manufacturing investments made by LG and Samsung, Park Gi-han noted that the timing of the investments 'confirms the hypothesis of a follow-the-leader internationalization strategy' (Park Gi-han 1996: 322). Looking at the fuller picture across western, central and eastern Europe, the hypothesis seems to hold true generally, with some notable exceptions. In the case of investment in sales facilities, the gap between the leading investment and the second investment was between one and three years in nine out of the 14 host countries. There was one gap of six years and, in four countries, no reaction at all to the investment by the other two firms. In terms of production, there was a swift reaction (within one to two years) in three out of 12 cases, one lag of seven years, and eight cases where the initial investment provoked no reaction from the other two rival firms.

Out of the 14 cases of sales investment in Europe, Samsung acted as leader in six countries (Portugal, the United Kingdom, France, Spain, Italy, and Sweden), spearheading the Korean advance into western Europe. This may reflect Samsung's reaction to LG's early advance into and rapid dominance of the US market. LG took the initiative in just two countries: Germany (the first case of Korean investment in western Europe), and Hungary. Daewoo was the pioneer in investing in central and eastern Europe, setting up sales subsidiaries in five former communist bloc countries between 1993 and 1996.

Only four countries in Europe hosted sales investments by all three firms and all four are located in western Europe – Germany, the United Kingdom, France and Italy. LG led the advance into Germany in 1980; although Samsung was quick to establish a sales subsidiary two years later, it was more than a decade after the initial investment was made that Daewoo entered the market. Samsung was followed by both its rivals into the United Kingdom (LG two years later and Daewoo seven years after that), France (LG after two years, followed by Daewoo a year later), and Italy (a gap of three years for Daewoo and then two years for LG).

The six countries that hosted sales investments from two out of the three firms are evenly split between western and central/eastern Europe. Samsung's moves into Spain and Sweden prompted reactions from Daewoo (within two years and six years, respectively), and Daewoo invested in the Hungarian market within a year of LG's advance into the country. Three of Daewoo's investments were matched by Samsung: the Netherlands within one year, Russia in two, and Poland in three. However, neither LG nor Samsung saw fit to follow Daewoo into Romania, the Ukraine or the Czech Republic, and Samsung was the only one of the Big Three to establish a sales facility in Portugal.

The only country in which all three consumer electronics firms established a production facility is the United Kingdom, with only a year

Table 8.2 Oligopolistic Reaction FDI in Europe by the 'Big Three'[a], June 1997.

	Sales 1	*Sales 2*	*Sales 3*	*Gap*	*Prod. 1*	*Prod. 2*	*Prod. 3*	*Gap*
Germany	1980 LG	1982 SS	1992 DW	2/10	1986 LG			NR
Portugal	1982 SS			NR	1982 SS**			NR
U.K.	1984 SS	1986 LG	1993 DW	2/7	1987 SS	1988 LG	1988 DW	1/0
Turkey					1987 LG**	1989 SS**		2
France	1988 SS	1990 LG	1991 DW	2/1	1988 DW			NR
Spain	1990 SS	1992/6 DW		2	1989 SS	1996 DW		7
Italy	1991 SS	1994 DW	1996 LG	3/2	1989 SS**	1990 LG**		1
Hungary	1992 LG	1993 DW		1	1989 SS			NR
Sweden	1992 SS	1998 DW		6				
Poland	1993 DW	1996 SS		3	1993 DW			NR
Russian Federation	1993 DW	1995 SS		2				
Czech Rep.	1993 DW			NR				
Uzbekistan					1993 DW			NR
Neth'lands	1994 DW	1995 SS		1	1996 DW			NR
Kazakhstan					1994 DW			NR
Romania	1995 DW			NR				
Ukraine	1996 DW			NR				

a Samsung (SS), Daewoo (DW) and LG
** Joint venture
NR No reaction from competitors.
Sources: EIAK 1997f, Park Gi-han 1996: 321, and Jun and Simon 1992: 196.

separating all three investments. This reflects the strong preference for the United Kingdom among Korean investors in Europe. (Empirical evidence [Park Gi-han 1996 *et* al] suggests that factors including the use of the English language, government incentives, good infrastructure and access to the continental European markets acted as strong incentives for investment in the UK.) Samsung's advances into Italy and Spain were matched by LG within one year and Daewoo in seven years; LG's decision to manufacture in Turkey prompted a response from Samsung two years later. In eight of twelve European countries that hosted Korean investment in local manufacturing, there was only one Korean firm active in production: LG in Germany, Samsung in Portugal and Hungary, and Daewoo in France, the Netherlands, Poland, Uzbekistan, and Kazakhstan.

Thus it is apparent that, while there is strong evidence of oligopolistic reaction among the Big Three in terms of their investments in *sales* subsidiaries in western Europe (a reaction could be observed in eight out of the 14 countries in which investments were made), the evidence in terms of investment for *production* is somewhat weaker (a reaction in four countries out of 12). However, there was no reaction from the other two companies to sales investments made by Samsung in Portugal and by Daewoo in the Czech Republic, Romania and the Ukraine. Similarly, there was no reaction to LG's manufacturing investment in Germany, Samsung's investments in Portugal and Hungary, and Daewoo's entry into France, Poland, Uzbekistan, the Netherlands, and Kazakhstan.

A second interesting point arising from this analysis is that, while oligopolistic reaction theory appears to have some validity in explaining *reverse* investment by Korean consumer electronics companies, there seems to have been little or no oligopolistic reaction to investments made in developing countries lower down the economic pecking order. In the case of investment in eastern Europe, out of 10 cases of pioneering investment in either sales or manufacturing, only three resulted in a similar move by a rival. Daewoo followed LG into Hungary (sales), and Samsung followed Daewoo into Poland and Russia (also sales). In the remaining seven cases, there was no reaction to the original investment. In contrast, in the case of western Europe, out of 16 cases of investment in sales or production, the initial move was followed by both rivals on five occasions and, in six cases, by one rival firm. There was no reaction from rival firms to the investments made by Samsung in Portugal (sales and production), LG in Germany (production), and Daewoo in France and the Netherlands (production).

A further point to note is that, in all four cases where the time lag between investments was greater than three years, the investor was Daewoo. Furthermore, out of 12 cases where an oligopolistic reaction was not observed, six were investments made by Daewoo. In almost every case where investments were made by both LG and Samsung, the time lag was short: between one and two years in six cases, and five years in Italy only. Out of the nine western European countries where Korean investment took place, LG and Samsung were both active in sales and/or production in five (Germany, the United Kingdom, Turkey, France and Italy). Samsung was the only firm to invest in Portugal, was followed by Daewoo into Spain and Sweden, and itself followed Daewoo into the Netherlands.

As far as patterns of investment are concerned, it would appear that it would be more accurate to characterize FDI patterns as resembling a game of 'leapfrog' rather than 'follow-the-leader', as each of the 'Big Three' has taken it in turns to lead the advance into new host countries. Although Daewoo was perhaps perceived as the third ranking firm out of the 'Big Three' in terms of domestic and global market positions, Daewoo ranked with Samsung in terms of pioneering FDI in Europe.[2]

Samsung was the first Korean electronics company to set up a sales subsidiary in Portugal, the United Kingdom, France, Spain, Italy, and Sweden, while Daewoo led the advance into Poland, Russia, the Czech Republic, Romania, the Ukraine and the Netherlands. LG was the pioneer in just two countries: Germany and Hungary. In terms of manufacturing facilities, Samsung was the first Korean investor in Portugal, the United Kingdom, Spain, Italy and Hungary. LG led the advance into Germany and Turkey, while Daewoo was the first to move into Poland, Uzbekistan, Kazakhstan, and France. Broadly speaking, in terms of both a sales presence and manufacturing facilities, Samsung and LG led the Koreans' advance into western Europe, with Daewoo breaking open the new markets in eastern Europe.

Three conclusions can be drawn from these facts: firstly that the real oligopolistic reaction occurred between Samsung and LG in their investments in western Europe. Secondly, while Daewoo followed its two rivals into the key markets of western Europe (the United Kingdom, France, Germany, and Italy), it did so at a much later stage. Thirdly, Samsung and LG chose to focus their efforts on western Europe, while Daewoo established a sales and manufacturing network in eastern Europe. This may have reflected Daewoo's desire not to compete with its rivals in western Europe, but to establish itself in a region where its rivals were not so active (central and eastern Europe).

These conclusions raise an important question which is beyond the scope of this study: why were LG and Samsung content to leave Daewoo to open up the eastern European markets? In June 1997, both LG and Samsung had sales offices in Hungary, and Samsung had established sales subsidiaries in Poland and Russia. Daewoo, on the other hand, was operating a network of six sales subsidiaries and three production facilities in eastern Europe. It may be that the more established companies felt that their network within western Europe was sufficient to service the needs of the market. Equally, they may have taken the view that direct investment in the countries of central and eastern Europe was best left until a greater degree of economic and political stability had been achieved. Whatever the reasoning behind the decision, it is clear that neither Samsung nor LG felt that the oligopolistic *status quo* would be threatened by these moves.

The data presented in this section has shown that oligopolistic reaction theory can only offer a partial explanation of the flows of Korean investment in central and eastern Europe. In these regions, the Koreans were investing in countries where they could combine their comparative advantages of technology, production experience and management expertise with the locational advantages of low-cost land and labour, and access to the key export markets in western Europe. The same holds true for those countries in western Europe which offered the same advantages to the Koreans, such as Portugal.

Interestingly enough, however, oligopolistic reaction theory appeared to be more applicable to the cases of RDI in the industrialized countries of western Europe, where the Koreans' advantages were weak or non-existent. This may have been due to a perception that the industrialized countries of western Europe were the more important export markets within Europe at the time, but it was almost certainly linked with these countries' membership of the European Union, and the Koreans' desire not to be locked out of 'Fortress Europe'.

However compelling the evidence for 'follow-the-leader' or 'leapfrog' patterns of investment may be, it is the author's belief that oligopolistic reaction theory must be integrated into a broader framework in order to explain why Korean consumer electronics firms have engaged in reverse direct investment in the absence of a clear firm-specific monopolistic advantage.

KOREAN CONSUMER ELECTRONICS INVESTMENT IN EUROPE: A THEORETICAL/MACROECONOMIC PERSPECTIVE

Over the past two decades, studies of Korean foreign direct investment have tended to focus on micro-level issues such as individual firms' motivations for investment, their competitive or comparative advantage, and the success or failure of their venture. It is the intention of this study to take a broader view, taking a step back and considering the key macro-level elements that influenced the explosion in FDI activity by Korean firms witnessed between 1987 and 1996. In doing this, we will draw on and expand upon the work carried out by Yun Tong-jin and Jun Yongwook. Specifically, we will look at Jun's idea of 'structural irreversibility' and Yun's concept of 'involuntary internationalization', considering the impact of changes in the domestic and global business environment, the influence of the characteristics of the consumer electronics industry, and the effect of ROK government policy, both economic and diplomatic.

The business environment, 1963–1986

The economic development policy adopted by successive Korean governments during the 1960s and 1970s was one of promoting export-led industrialization, with financial and administrative support focused to a great extent on a small number of large and highly diversified conglomerates (the *chaebŏl*). Government and big business formed a strong and successful alliance, in which enterprises that followed the development path indicated by the government were given preferential access to funding and business opportunities, and were protected, by and large, from competition with foreign companies in the domestic market. These favoured companies were encouraged to grow rapidly and diversify quickly, focusing on quantitative rather qualitative growth.

In addition to the financial and administrative support received from the government, the conglomerates benefited from strict labour laws that effectively curbed worker unrest and helped keep wages low. The government also endeavoured to protect Korea's growing industries by restricting access to the domestic market for foreign goods and services. Foreign participation in the economy was permitted on condition that it brought with it the capital and technology needed for Korea's rapid industrialization.

The government's policy was, to a great extent, shaped by the characteristics of the Republic of Korea, most notably the paucity of natural resources, the lagging levels of technology accumulation, the lack of investment capital, and the small size of the domestic market. In order to achieve the economies of scale required to secure price competitiveness, industrial facilities had to be built with capacities far in excess of the requirements of the local market. Exporting the surplus brought in the foreign exchange needed to acquire the raw materials, intermediate and capital goods required to support rapid economic development.

In adopting an export-led industrialization policy, the government was exploiting the one advantage possessed by Korea at this stage of its economic development, namely its well-educated, low-cost workforce. The drive for development was also supported by the strong work ethic, desire for education, and willingness to work for the good of the group, which were essential elements of the Confucian social system that continued to exert its influence over Korean society in the late 20th century. An external factor that cannot be overlooked is the threat from North Korea; the policy of rapid industrialization adopted by the Park government in the 1960s and 1970s aimed to create a strong state capable of defending itself against the threat of communism.

As can be seen in Figure 8.1, between 1963 and 1986, rapid growth in the Korean economy and strong export performance were fostered by a favourable domestic and global business environment. Korean manufacturers and exporters enjoyed a strong competitive advantage that derived principally from their access to a large pool of high-quality and low-cost labour. Manufacturers used mature technology imported from industrialized countries to mass produce low value-added, low-tech, labour-intensive goods. Mass production enabled the Koreans to achieve

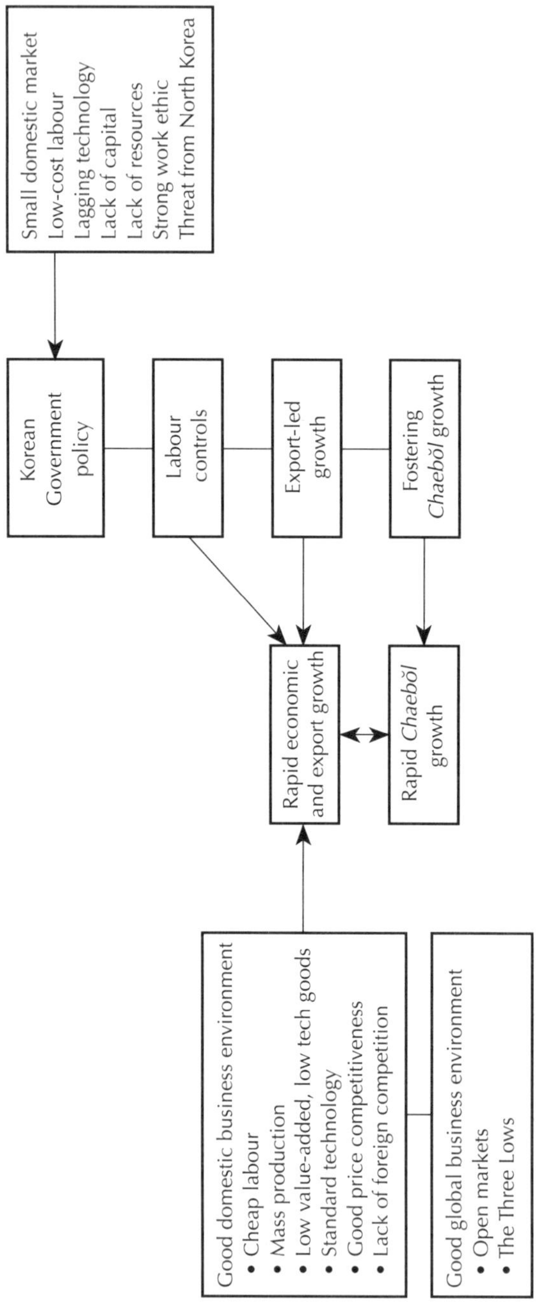

Figure 8.1 The Business Environment and ROK Government Policy, 1963–1986.

economies of scale and price competitiveness in world markets. A heavy reliance on OEM exports allowed the Koreans to focus their efforts on the assembly of imported parts, components and semi-finished goods, and sell them in a global market that was relatively unencumbered by trading restrictions.

From the middle of the 1980s, developments in the world economy contributed to Korea's stellar economic performance; low international interest rates reduced the burden of debt repayments, low oil prices eased the pressure on production costs, and favourable exchange rate movements enhanced the price competitiveness of Korean goods in the global market.

The business environment, 1987–1996

As we have seen in Chapter 5, the majority of Korea's foreign direct investment occurred between 1987 and 1996. Out of a net cumulative total of US$13.8 billion at the end of 1996, more than 95 per cent in value terms (US$13.1 billion) and 93 per cent in volume terms had been invested during this ten-year period. It is now clear from our understanding of the developments in the Korean economy, the consumer electronics industry, and the global business environment, that this surge in investment activity was due to a complex combination of domestic and international factors.

Figure 8.2 shows the influence of these factors on patterns of Korean FDI between 1987 and 1996. The continuing growth of the Korean economy in general and export performance in particular was threatened by changes in the domestic and global business environment. In Korea, rising wages, land prices, and interest rates eroded the price competitiveness on which Korean exporters had relied in the past. The situation was exacerbated by the appreciation of the wŏn against major currencies, which made Korean products more expensive in global markets.

Manufacturers who had depended on an abundance of low-cost, high quality labour found that advantage slipping away with the high wage increases that followed the implementation of democratic reforms and the formation of effective trade unions in 1987. Between 1986 and 1995, nominal wages rose at an average annual growth rate of 15.3 per cent and by 1994 manufacturing wages in Korea were around 30 per cent higher than wages in Taiwan and Hong Kong (Wang 1997: 59). Along with the increase in labour unrest, employers found themselves confronted with labour shortages as Koreans became increasingly reluctant to do the 'Three D' jobs: work that was difficult, dirty or dangerous. As disposable incomes and standards of living rose, Korea's traditionally strong work ethic weakened, and there was a growing disinclination among Koreans to work long hours for the good of the company and nation.

Increases in the price of land – always a commodity in short supply given that some 70 per cent of the Korean peninsula is covered by mountains

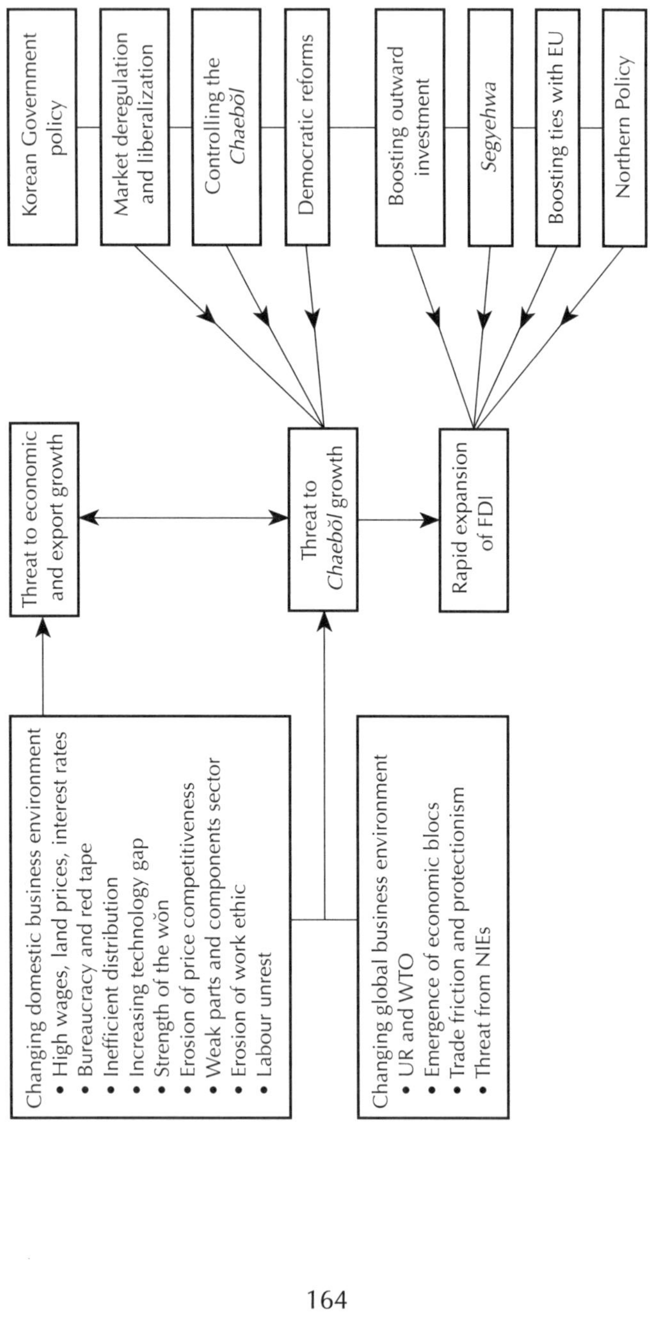

Figure 8.2 Changes in the Business Environment and ROK Government Policy, 1987–1996.

(Korea Overseas Information Service 1993: 7) – and higher domestic interest rates added to the financial burden for enterprises wishing to establish or expand production facilities. Insufficient investment in social overhead capital in the period of rapid economic growth had led to severe problems in transportation, warehousing and distribution, which further exacerbated the problems facing manufacturers and exporters.

Levels of investment had also been low in the area of research and development, leading to a widening of the technology gap between Korea and industrialized nations and a deepening of Korean firms' dependency on those nations for advanced technology, core parts and components. Reliance on a strategy of focusing on OEM sales (and thus allowing the foreign buyer to shoulder the burden of marketing, distribution and after-sales service) left the Koreans weak in key areas of the manufacture and sales process. Moreover, continuing government controls over and intervention in the economy, manifested in tight controls over the allocation of financial resources and high levels of red tape, acted as a deterrent to investment in the early 1990s.

The late 1980s and early 1990s also saw major changes in the global business environment which posed a threat to Korea's export performance and long-term economic growth. South Korea's very success in the late 1980s in terms of strong economic growth and rapid expansion of exports led to the imposition of import restrictions on Korean products and the erection of non-tariff barriers. According to a report published by UNCTAD, all Korea's major consumer electronics export items, including televisions, video cassette recorders, compact disc players, microwave ovens, and colour picture tubes, have faced protectionist measures in key markets (UNCTAD 1997: 32).

Moreover, the emergence of regional economic blocs such as the North American Free Trade Agreement (NAFTA) and the European Union (EU) posed a further threat to countries that were not members of any such regional grouping. As we have seen, the prospect of the creation of 'Fortress Europe' proved to be a powerful incentive for corporations to review their policy on Europe and take steps to expand or establish a presence within the EU ahead of the creation of a single market.

As Korean enterprises struggled to maintain their competitiveness in global markets in the face of rising production factor costs, there emerged a new threat in the form of developing countries and Newly Industrializing Economies that still enjoyed the price advantages now lost to the Koreans. Given their lack of focus on and investment in R&D, brand name development, and marketing in the 1970s and 1980s, ROK firms found themselves caught in an uncomfortable middle position, unable to compete on price in the low end of the market and yet unable to compete on quality in the high end of the market.

The conclusion of the Uruguay Round and the establishment of the World Trade Organization signified a global commitment to the creation of a free trading environment and an era of borderless competition. Countries like South Korea, whose governments had in the past taken measures to boost the performance of their domestic industries while restricting access to the local market, would now have to make a commitment to the full liberalization and deregulation of their markets and economic systems.

These changes in the domestic and external environment posed a major threat to the continuing growth of the South Korean economy in general and to the growth of the *chaebŏl*, who (until the mid-1980s) were acknowledged to be a major driving force behind Korea's rapid economic growth and export expansion. Even before the economic crisis of 1997, public sentiment regarding the *chaebŏl* had become increasingly negative, and successive governments had found themselves under pressure to curb the activities of the conglomerates. Attempts had been made to increase support for small- and medium-sized enterprises, restrict the *chaebŏls'* expansion, and encourage a streamlining of their business operations.

The past fifteen years have seen major changes in the environment in which the conglomerates operate. The ROK government's commitment to market deregulation and liberalization (made in the 1980s and 1990s as Korea moved towards its dream of joining the OECD and attaining advanced nation status), successive governments' efforts to control the *chaebŏl* in the face of growing criticism of their power and activities, and former president Kim Young-sam's vision of globalization (*segyehwa)* have all contributed to the creation of a new business environment for domestic enterprises.

Among the stated aims of *segyehwa* are the promotion of FDI and the liberalization and deregulation of the Korean economy, allowing greater access for foreign goods and services. This shift in government policy – from one of restricting imports and protecting Korean firms to one of liberalization – made an increase in foreign competition in the Korean market inevitable. This in itself proved to be a powerful incentive for the Korean conglomerates to develop their own global networks and take steps to acquire the advanced management skills and expertise that would enable them to compete successfully on a worldwide basis.

How have these changes influenced the patterns of Korean outward foreign direct investment in the late 1980s and 1990s? After 1987, Korean manufacturers whose main competitive advantage had lain in economies of scale achieved through mass production and low domestic wage levels found that advantage disappearing rapidly. For these firms, the principal means of survival was the relocation of their labour-intensive production facilities overseas, to countries which still enjoyed the advantage of low-cost labour. The facilities at home could then be upgraded to support the production of more technology- and capital-intensive goods. The deterioration of the business environment for labour-intensive manufacturing in Korea was, therefore, a major incentive for businesses to boost their FDI activities in developing countries.

It should be remembered that this deterioration in the business environment at home also affected the Korean companies that had invested in industrialized countries in the early 1980s. The ability to manufacture using cheap parts, components and semi-finished goods had enabled the investors to create a short-term competitive advantage in their overseas operations. The loss of that advantage did not, however, lead to a decline in FDI levels or the mass closure of foreign subsidiaries. Rather, the need to maintain and even expand a local presence in industrialized markets became more apparent as global business conditions changed for the worse.

A second incentive was the strengthening of import restrictions and non-tariff barriers against Korean products in industrialized countries and the emergence of regional economic blocs. Korean companies grew concerned that the acceleration of 'bloc-ism' would lead to measures designed to advance the interests of bloc members to the detriment of non-members. The recognition of the advantages to be gained from operating inside the bloc in terms of market access, avoidance of trading restrictions, and (in the case of industrialized countries) access to superior technology and management expertise prompted many companies to establish facilities within the newly emerging blocs. A prime example of this, of course, was the European Union; being inside the bloc would be vital if companies were to enjoy the benefits of economic integration.

Government policy was a third factor influencing the decision to move overseas; in the late 1980s, the Roh administration took steps to encourage outward FDI as a means of managing the swelling current account surplus and easing criticism and pressure from major trading partners. The opening up of the Korean market to foreign goods and services (partly as a result of this pressure) brought many Korean firms into direct competition with enterprises from industrialized countries for the first time. Given the ROK government's commitment to deregulation and liberalization and pressure from international bodies and trading partners to honour that commitment, Korean enterprises could no longer rely on protection from official sources.

Additionally, for the *chaebŏl*, it was becoming increasingly clear that the glory days of 'Korea Inc.' were waning; in the future there would be increasing pressure on them to restructure and reform. The government's commitment to playing a full role in the World Trade Organization and other international trade bodies would make it harder to justify the adoption of measures designed to protect Korean companies and the domestic market. Furthermore, the implementation of democratic reforms in the summer of 1987 led to sweeping changes in the labour market, including the formation of free and effective trade unions. The immediate result was a upsurge in industrial unrest and spiralling wage increases. Although the government would continue to try to bring pressure to bear on employers and workers to reach swift and amicable settlements, direct intervention and suppression of workers' rights would no longer be countenanced.

For all Korean manufacturers, a new competitive edge had to be created in areas other than price and, for many firms, this meant gaining access to advanced technology and management expertise available only through alliances with foreign enterprises. The Korean vision of *segyehwa* – world-class enterprises competing on a global basis through the establishment of international production, marketing and distribution networks – could not be achieved by remaining in the domestic market or maintaining only a nominal presence overseas.

Changes in government policy regarding international relations also encouraged Korean firms to move overseas. The establishment of diplomatic relations with central and eastern European nations in the late 1980s and early 1990s opened up potential host countries for Korean investment, offering low-cost labour, cheap land, and access to the European Union in the medium to long term. The development of closer links with the European Union itself, after decades of mutual neglect, encouraged Korean firms to turn their eyes away from the United States and Japan as sources of technology, marketing skills and management know-how, and look towards the world's largest market.

This shift in policy, towards a more multidimensional approach to diplomacy, may also be linked to developments in inter-Korean relations and changes in the relationship between the United States, North and South Korea in the early 1990s. Since the end of the Korean War, North Korea has pursued a policy of seeking direct talks with and recognition by the United States, with the primary aim of replacing the 1953 Armistice Agreement with a permanent peace treaty, and having US forces withdraw from the Korean peninsula. This strategy was seen by many South Koreans as an attempt to drive a wedge between themselves and their closest allies (Ahn 1986: 284–285), and concerns were heightened in the early 1990s by the US government's adoption of a policy aimed at developing political and economic contacts with the DPRK.

Beginning with the implementation of the Bush administration's 'Modest Initiative' in 1988, which provided for limited trade, diplomatic and non-governmental contacts with North Korea, and culminating in the 1994 'Agreed Framework between the United States and the DPRK', which brought to an end the 1993–1994 nuclear crisis, there have been clear signs of the US government's desire to engage with rather than confront the North Koreans.[3] Among the key provisions of the Agreed Framework were assistance for North Korea's nuclear energy programme, the partial lifting of US-imposed trade sanctions, the opening of liaison offices in the DPRK and the United States, and the eventual establishment of full political and economic relations between the two nations (Independent Task Force on Managing Change on the Korean Peninsula 1998: 7–8).

The US government consistently offered its South Korean allies reassurances that any expansion of contacts with the north would occur in consultation with the ROK and in parallel with improvements in north-south relations. Despite such reassurances and, indeed, despite South Korea's own efforts to engage with her northern neighbours, these developments may well have provided additional impetus for the ROK government to seek closer links with and greater diplomatic support from the European Union.

Another important factor in the FDI activities of Korean consumer electronics firms is the development of the industry itself. The impact of the Korean government's policy regarding the consumer electronics industry set off a chain reaction which ultimately led to FDI as a simple matter of survival. As we have seen in Chapter 4, the government implemented plans and introduced legislation aimed at developing the consumer electronics industry into a strategic export industry capable of driving the nation's exports and its overall economic performance.

These measures included the designation of the consumer electronics sector as a strategic export sector and an industry for accelerated development in 1966, and the promulgation of the Electronics Industry Promotion Law (1969), the Long-Term Plan for the Enhancement of the Electronics Industry (1982), and a five year-plan to develop core technology, parts and components (1992). These major items of legislation were supported by additional measures such as the 1961 ban on imported radios (while steps were taken to increase the diffusion of radios in farming and fishing villages), and the 1974 decision to ban the sales of colour television sets in South Korea.

The government supplied the financial and administrative support for the creation of massive production facilities through which Korean firms could achieve economies of scale and price competitiveness. When the goods were produced, there was insufficient demand in the domestic market due to the large volume of supply, low levels of income, and government restrictions. Firms were, therefore, forced to export by local conditions and encouraged to export by government policy.

Government legislation not specifically targeting the consumer electronics industry has also boosted FDI; the easing of regulations on outward FDI as a means of controlling the current account surplus in the latter part of the 1980s and thus easing friction with major trading partners has already been noted. In addition, the relaxation of restrictions on overseas financing enabled large firms to raise funds for FDI projects at a lower cost than in the domestic financial markets, through the issue of financial instruments including convertible bonds and floating rate notes.

The characteristics of the consumer electronics industry have also played their part; the high level of export dependency and the strategy of *chiptchung houshik such'ul* (flooding industrialized country markets with price competitive goods) resulted in trade friction and left Korea vulnerable to changes in the global market. The strong foreign presence in production and exports led to the acquisition of standard technology from foreign investors in Korea in the early stages of development, and dependence on OEM exports encouraged low levels of R&D, a lack of focus on marketing, and a concentration on the production of undifferentiated goods.

The lagging technology levels and weak parts and components sector (due to over-emphasis on the quantitative expansion of simple assembly operations) also added to the need for FDI in the 1990s, as a means of acquiring technology. Finally, the creation of an oligopolistic market structure due to the government's policy of favouring the *chaebŏl* and encouraging market domination by a few, large-scale producers encouraged 'follow-the-leader' or 'leapfrog' patterns of foreign direct investment by Korean multinationals in overseas markets.

Summary and Conclusions

The past decade has seen a rapid increase in foreign direct investment activity by Korean corporations in general, and by firms from the consumer electronics sector in particular. The late 1980s and early 1990s saw a strong advance into the developing and industrialized countries of western, central and eastern Europe by the Korean consumer electronics giants, Samsung, LG and Daewoo. Responding to changes in the domestic and global business environment, these Korean manufacturers sought low-cost production and export bases in developing and less developed countries, while seeking to protect key export markets and acquire the technology, management skills and expertise needed for globalization through investment in the industrialized nations of Europe. The internationalization process of the consumer electronics industry with regard to Europe was unusual in that the first advances into the region were made into industrialized countries, with large-scale investment in developing countries following at a later date.

The increase in investment activity gave rise to a debate in Korea on the ability of existing FDI theories to explain in full the forms and flows of Korean overseas investment. The theories developed by Hymer, Dunning, Buckley and Casson *et al* took as a basic premise the possession of a firm-specific monopolistic advantage that could offset the cost of foreignness deriving from operating overseas, and allow the investor to compete on an equal footing with local firms and other multinational enterprises. However, the consensus among Korean scholars was that such ownership advantages as Korean investors possessed were too weak to offset the cost of foreignness and were likely to be only temporary in nature. Therefore, within the OLI (ownership-location-internalization) paradigm of Dunning's Eclectic Theory, Korean firms appeared only to possess temporary and comparative internalization advantages in the latter half of the 1980s (deriving from low labour costs back in Korea which could be internalized through the supply of parts, components and semi-finished goods to the overseas subsidiary) and to benefit from the location-specific advantages offered by host nations.

As Korean companies moved to establish sales and production facilities in industrialized markets, Korean scholars concluded that a new theoretical approach was needed to explain the phenomenon of reverse direct investment. Some scholars developed an approach based on, but extending, the western theories of FDI, while others focused their efforts on developing an independent theory of RDI. This study has presented for the first time in English a survey of Korean approaches to foreign direct investment theory in general, and the works of three scholars in particular. Having analyzed the works of Wŏn Chong-gŭn, Yun Tong-jin and Jun Yongwook, the study focused on two aspects of their work as a basis for developing a broad macroeconomic framework within which Korean investment in Europe can be understood.

The first aspect was the application of oligopolistic reaction theory to Korean investment in Europe; Wŏn (1986) incorporated the theory into his explanation of RDI, and other scholars including Jun (1988) and Park (1996) have commented on the impact of the oligopolistic structure of the consumer electronics industry on Korean FDI activity. This study, which has covered a broader field than previous works by incorporating data on western, central and eastern Europe through to the end of 1996, revealed evidence of oligopolistic reaction among the 'Big Three' consumer electronics companies. The 'leapfrog' pattern was strongest between Samsung Electronics and LG (with the two companies taking it in turn to lead the way) in western Europe, and in the sales sector. While Daewoo Electronics followed its rivals' lead into major industrialized countries, it also took steps to create a sales and manufacturing network in central and eastern Europe. However, these investments provoked little reaction from Daewoo's competitors.

The second aspect of the Korean debate on which this study has focused is the consensus that FDI activity was, to a great extent, inevitable due to the structural characteristics of the Korean economy and the consumer electronics industry. Writing in 1986, Wŏn Chong-gŭn commented that the Korean firms were 'structurally forced to rely on overseas markets' (Wŏn 1986: 129); Jun Yongwook incorporated the concept of 'structural irreversibility' into his Structural Market Defence Theory (Jun 1989: 182), and Yun Tong-jin referred to reverse direct investment by consumer electronics firms as a case of 'involuntary internationalization' (Yun 1993: 242).

Following a detailed overview of the development and structure of the Korean consumer electronics industry, patterns of Korean investment in Europe were analyzed in the light of the economic development path followed by the Republic of Korea and the growth and characteristics of the consumer electronics industry. It was seen that a wide variety of factors, both domestic and global, had combined over the past decade to create a situation in which overseas investment was inevitable, even when theoretically unfeasible – a case of 'involuntary internationalization'.

The economic policies adopted by successive governments and the development path followed by the Korean consumer electronics industries set in motion sequences of events which led inexorably to direct investment overseas. Looking at one such chain of events, we can see that, under an export-led growth policy in which companies are encouraged to operate large-scale plants with output in excess of domestic demand in order to achieve price competitiveness, and with a corporate and industrial policy of focusing on the export of a few mature products to a limited number of developed markets, the result will be the flooding of industrialized markets with low-price goods. Success will lead to the imposition of trade restrictions in these key markets and the very real prospect of a decline in export performance or even the loss of the market. Manufacturers facing the prospect of declining exports and falling profitability, but lacking the time to develop alternative markets, will have little choice but to make a direct investment overseas as a means of securing the market and ensuring the growth, and even the very survival, of the firm.

Under another scenario, a company which pursues a rapid-growth strategy in line with government policy and relies on OEM sales as a convenient route to rapid export growth, pays scant attention to developing technology, enhancing product quality, establishing a brand name or creating marketing and distribution and networks. The availability of low-cost labour, government support for employers, the weakness of the trade union movement, and the compliance of the work force provide a seemingly sound basis for maintaining price competitiveness. However, the implementation of political reforms and the establishment of free and effective trade unions lead to labour unrest and wage increases far in excess of productivity gains, which erode the firm's price competitiveness. Meanwhile, better living standards, the development of the service industry, and rising wage levels lead to an erosion of the traditional work ethic and the creation of labour shortages. With no established reputation for quality, the company must find a low-cost production base or lose its competitive edge. Once again, foreign direct investment (this time in developing and less developed countries) is the route to survival.

Finally, for the company engaged in intense oligopolistic competition within the domestic market, a move by one of its rivals to set up a sales office or manufacturing base overseas could pose a threat to the *status quo* at home. The pioneer investor may gain an advantage by developing close contacts with suppliers and clients in the market. Additionally, the company may gain prestige from an image as a global player manufacturing and selling internationally-recognized products, and may benefit from the approval of the home government, should official policy be directed towards the expansion of outward investment. The company left behind has just two choices: to remain at home and risk losing ground in the competition with its rival, or to match the move and attempt to restore the oligopolistic balance.

Although, as these examples show, global and domestic factors combined with the structural characteristics of the Korean economy and the consumer electronics industry have led to 'involuntary internationalization' by companies who might otherwise have preferred to continue operating in and exporting from their home market, there were other developments that encouraged rather than forced FDI. Many of these factors fall into the category of location-specific advantages, including as they do the low cost of labour and land in eastern European countries, incentives offered to foreign investors by European governments, opportunities for participation in the privatization of state-run enterprises, and so forth. However, the 'pull' factors for FDI have also been affected by changes in ROK government policy. The recognition of the importance of the European Union and the promotion of closer ties with the world's largest single market, and the establishment of diplomatic ties with the countries of central and eastern Europe, have afforded significant new opportunities for would-be Korean investors in the past decade.

This study has analyzed patterns of Korean FDI from 1968 through to the end of 1996. The decision to focus the analysis on this period was a reflection of the author's belief that the economic crisis of 1997 and its aftermath will have far-reaching effects on current and future overseas investment by Korean corporations. In May 1999, it was reported that Korean enterprises had scaled back their overseas investment activities in the wake of the crisis. Data published by the Ministry of Finance and Economy revealed that between 1996 and 1998, Korean FDI had fallen by 17.9 per cent to US$5.1 billion on a total permitted basis.[1] The data also showed that FDI had fallen even more sharply in the first two months of 1999, when the year-on-year rate of decline accelerated to 28 per cent (Kwak 1999: 26).

In the light of these developments, an important area for future research will be the medium- and long-term impact of the 'IMF Crisis' on Korean FDI. Will the corporate restructuring begun in late 1997 lead to a retrenchment in new FDI activity and the scaling down or closure of projects already in existence? The answers to this question could have important implications for countries in which the Koreans have already established sizeable manufacturing facilities. Although the Korean consumer electronics companies have been major investors overseas, firms from other industries have also relocated their production facilities overseas as part of their globalization strategy. A further area for study would be the analysis of FDI patterns in, for example, the automobile industry, to determine whether the patterns identified in this study are common to other sectors of Korean industry.

The four questions posed at the beginning of this book have now been answered. It is clear that, in the early 1980s, Korean consumer electronics companies did invest in the industrialized countries of western Europe in the absence of a strong firm-specific monopolistic advantage. The temporary advantage they were able to create was easily eroded by changes external to the firm and was not replaced by a stronger, more lasting advantage. In contrast, Korean firms investing in the countries of central and eastern Europe have done so on the basis of the advantage they enjoy in terms of appropriate technology and experience of manufacturing in developing countries.

Given the lagging levels of technology that still exist in key areas of Korean industry (most notably in the area of core parts and components) and the failure of the Korean consumer electronics companies to invest heavily in research and development, it was not feasible that Korean RDI should take place on the basis of innovation, as Vernon suggested. Previous micro-level studies have analyzed the importance of country-specific locational advantage as 'pull' factors for investment. From a macroeconomic perspective, the locational advantages of the EU as the largest single market in the world and the countries of central and eastern Europe in terms of their own potential and the access they afford to the EU have acted as powerful incentives for Korean investors. Finally, an analysis of FDI data has revealed strong evidence of oligopolistic reaction in the countries of western Europe, as the two largest consumer electronics companies have endeavoured to defend this key market.

It is the author's hope that this study will make a significant contribution to the pool of knowledge relating to Korea – in terms of the Korean economy, Korean foreign direct investment, the consumer electronics industry, and Korea's relations with Europe. This study has drawn upon a broad range of Korean language materials to provide essential background information in a number of areas, creating a context within which further discussion can take place. It is hoped that the presentation of work by Korean scholars in the field of foreign direct investment will open up new areas of discussion for their western counterparts, and that the theoretical/ macroeconomic framework presented here will prove a valuable tool in understanding the patterns of and motivations for Korean foreign direct investment.

Postscript: The *Chaebŏl* and the Korean Economy in the 'IMF era'

In December 1997, the International Monetary Fund (IMF) reached an agreement with the government of the Republic of Korea, under which the IMF would provide the largest bailout package in its history and oversee a programme of reform and restructuring. The IMF would provide US$21 billion in stand-by credits over three years, while the World Bank and the Asia Development Bank would supply US$10 billion and US$4 billion respectively. A dozen of the world's industrialized nations agreed to provide a further US$20 billion in assistance (Suh 1998: 35). In return for the financial assistance, the South Korean government made a commitment to implement a programme of reform comprising:

> ... strengthened fiscal and monetary policies, far-reaching financial reforms and further liberalization of trade and capital flows, as well as improvement in the structure and governance of Korean corporations (*Business Korea,* March 1998: 41).

The short-term goals of the IMF programme were the attainment of 3 per cent economic growth in 1998, a reduction in the current account deficit to 1 per cent of gross domestic product (GDP), and curbing of inflation at or below 5 per cent at the consumer level. The policies also aimed to restore the value of the wŏn, maintain high interest rates to attract foreign investment in the domestic capital markets, implement and operate a flexible exchange rate system, and build up the nation's foreign exchange reserves. The incremental release of funding from the IMF would be dependent on the successful implementation of the proposed reform measures (Kim Chong-tae 1997a: 19 and Kwon 2000: 33).

Having agreed to the terms of the bailout package, president-elect Kim Dae-jung promised the Korean people that he would end the IMF's 'control of the economy' within 18 months, by repaying the funds lent to his government (*Business Korea* January 1998: 18). The approach adopted by the new Korean government was described as a 'four plus one' approach. The four areas of reform were the financial, corporate, and public sectors and the labour market; the 'plus one' referred to the government's commitment to the continuing liberalization of the Korean markets in general, and the financial markets and foreign direct investment activity in particular.

THE FINANCIAL SECTOR

Once the IMF agreement had been formalized, the Korean government moved quickly to stabilize the nation's financial markets. Laws were passed in the National Assembly, enhancing the autonomy of the central bank and bringing all the supervisory functions within the financial sector together under the control of one autonomous organization: the Financial Supervisory Commission (FSC) (Kwon 2000: 34). Thus the government ushered in a new era of more stringent reviews of capital adequacy and risk assessment, and the implementation of stricter standards of accountancy and disclosure.[1] An issue of government bonds raised a total of W64.0 trillion (US$53.3 billion)[2] for bailing out financial institutions by buying bad loans or providing injections of capital, and the government established the Korea Asset Management Corporation (KAMCO) to buy non-performing loans from the nation's commercial and merchant banks (http://www.economist.com/archive/ *Economist*, 10 July 1999).

Five insolvent commercial banks were closed down in July 1998, with their assets and liabilities taken over by five healthier institutions.[3] Nine other insolvent banks escaped closure on condition that they enter workout programmes, under which they would implement measures to strengthen their financial position and restructure their management. The two banks experiencing the greatest difficulties – Seoul Bank and Korea First Bank – were to be recapitalized and sold off to foreign investors.[4]

The government's offer of financial incentives for mergers of leading commercial banks bore fruit in July 1998, when Hanil Bank and the Commercial Bank of Korea agreed to merge their businesses. A report in the *Economist* estimated the two banks' bad loans at W14.8 trillion (US$12.3 billion) and revealed that the government would supply most of the W10 trillion (US$8.3 billion) needed to raise the new bank's capital adequacy ratio to acceptable levels. The same report noted that banks demonstrating reluctance to enter into merger agreements were likely to find the government less willing to purchase their bad loans (http://www.economist.com/archive/ *Economist* 8 August 1998). In 1997, the country's merchant banks, which had specialized in providing short-term finance to companies, held problem loans valued at US$8.8 billion (*Business Korea* September 1997: 24–25). Early in 1998, 10 insolvent merchant banks were ordered to close down, with the remainder placed under surveillance for review at a later date.

By the end of 1999, the government had spent a total of US$17.1 billion on buying non-performing loans and US$36.2 billion on recapitalizing financial institutions (*Business Korea* February 2000: 36–37). However, concerns persisted about the problem of non-performing loans; with 70 companies in workout programmes as of April 2000, failure to restore these firms to health could lead to further bankruptcies and an escalation in bad loans, as firms defaulted on repayments to the banks (http://www.economist.com/archive/ *Economist* 1 April 2000).

In the summer of 1999, the government turned its attention to the non-banking financial sector, including investment trust companies, insurance companies and securities houses. A key aim of action in this area was to prevent the conglomerates using these firms as a source of finance. In 1999, the top five *chaebŏl* controlled 34.7 per cent of the non-banking financial sector, with shares of more than 50 per cent in the securities, insurance and credit card businesses (Kim Seung-chuan 1999: 39). Under the new system, non-banking financial institutions would be required to appoint outside directors to their boards, subject their activities to the scrutiny of an internal compliance officer, and enhance the powers and rights of minority shareholders. Investment trust companies would be required to undergo an audit of large-scale asset management funds by outside accounting firms, and disclose their investment and trust assets. Listed firms would have to tighten their accounting standards and prepare consolidated financial statements, while unlisted companies would be obliged to publish business reports on a quarterly basis (*ROK Economic Bulletin* September 1999: 24–25).

In terms of foreign participation in the financial and capital markets, barriers to entry to the banking sector were lifted in March 1998, (with a few 'nominal' restrictions) with the aim of reducing government intervention in the markets and enhancing the competitiveness of domestic institutions (Kwon 2000: 34). The ceiling on stock ownership by foreigners was to be raised from 26 per cent to 55 per cent by the end of 1998, before being abolished completely (Kim Chong-tae 1997a: 19), and restrictions on investment in the bond market and the purchase of short-term money market instruments were also to be lifted (Kwon 2000: 37).

The implementation of a fully floating exchange rate system was brought forward from 2000 to December 1997 in order to curb fluctuations in the exchange rate and help increase Korea's holdings of foreign reserves (Park 2000: 64). On 1 April 1999, foreign exchange transactions were completely liberalized, eliminating all controls on capital movements by Korean and foreign companies and financial institutions. The second phase of reform (due to be completed by the end of 2000) would see the removal of restrictions on transactions by individuals, the abolition of limits on personal overseas travel expenses and the transfer of assets overseas by migrating families, and the complete liberalization of borrowing from overseas financial institutions (Kim Citi 1999c: 11).

THE CORPORATE SECTOR

In 1996, the combined debt of the top 30 conglomerates stood at US$192 billion and their average debt-to-equity ratio was 386.5 per cent. A year later, their debts had soared to US$255 billion, with a debt-to-equity ratio of 518.9 per cent and combined net losses of US$3.1 billion in 1997. In that year, five of the leading *chaebŏl*, holding assets valued at US$23.3 trillion and employing 107,000 workers, filed for bankruptcy (Lieberman and Mako 2000: 13 and *Business Korea* May 1998: 25). The collapse of these conglomerates highlighted the structural problems underlying their remarkable growth since the 1960s: rapid expansion funded by bank borrowings rather than internal reserves, leading to high levels of debt and vulnerability to changes in global export and financial markets.

As the first conglomerates filed for bankruptcy, others announced plans to restructure their businesses and improve their financial situation by selling assets, closing down marginal businesses and attracting fresh injections of capital (Sohn Young-ju 1997: 22–23). However, it was clear from the very beginning of the reform process that the new government would play a decisive role in promoting corporate reform and determining the future of Korea's business groups.

On 6 February 1998, the government and heads of the leading *chaebŏl* announced that they had successfully negotiated a five-point agreement on corporate restructuring. The reform of the corporate sector was to focus on the areas of ownership, corporate governance, and business, financial and management structures. Companies would be required to appoint directors from outside the firm (and, by implication, outside the controlling family) to their boards of directors, and increase the powers and rights of minority shareholders.

In a move to enhance the transparency of financial structures and accounting systems, all business groups would be required to prepare consolidated financial statements and abolish cross payment guarantees. (In the past, guarantees of one subsidiary's loans by another had enabled the *chaebŏl* to use successful affiliates to finance non-profitable companies within the group.) To improve their financial health, firms would have to take steps to reduce their debt-to-equity ratio, place a greater emphasis on profits and cash flow, and adopt a more conservative approach to their investment activity. In terms of their management structure, the conglomerates would be encouraged to focus on core businesses, enter into cooperative alliances with SMEs and foreign firms, and move away from their traditional paternalistic, fleet-style system of management towards the professional management of independent companies (Jung 1999: 84 and Yoo 1998: 11).

The approach to reform differed according to the size of the conglomerates. The top five were to formulate their own plans with the approval of their creditor banks, while the business groups ranked 6 to 64 were to enter into workout programmes supervised by their banks and the Corporate Restructuring Control Commission. SMEs were to have their debts rolled over for a year, while creditor banks worked to restructure their debts (Lieberman and Mako 2000: 15–17).

In January and May 1998, the chairmen of the top five conglomerates announced restructuring plans, promising the sale of subsidiary companies, the selection of core businesses, and a reduction in debt-to-equity levels (Kim Chong-tae 1998i: 12–13). Despite this action, the government instructed creditor banks and the newly-established watchdog, the Financial Supervisory Commission, to prepare a list of non-viable firms and companies eligible for workout programmes. In June, the FSC published a list of 52 firms targeted for closure, including 20 affiliates of the top five *chaebŏl*. However, the action was greeted with scepticism by observers, who suspected that the decision to close firms already sidelined by the conglomerates was little more than a public relations exercise for the benefit of potential foreign investors (*Business Korea* July 1998: 17–18).

Despite President Kim's declared intention to pursue democracy and a market economy in parallel – the very foundation of his approach to economic development, dubbed 'Djnomics' – there were signs from the very earliest days that the government would take an active role in the reform of the corporate sector. The 'Big Deals', announced in October 1998, provide an excellent example of the government-directed push for reform. By first encouraging and then pressuring the larger conglomerates to focus on their core areas of competence and merge or swap non-core divisions within their groups with affiliates of other *chaebŏl*, the government hoped to reduce excessive and overlapping investment and trim overcapacity in key industrial sectors[5] such as electronics and automobiles.[6] 'Djnomics' clearly did not involve allowing market forces to weed out the weaker firms and restructure Korean industry.

While moving to curb the power of the conglomerates, the government also took steps to promote the interests of the SMEs, whose development had been adversely affected by the rise and fall of the *chaebŏl*. In 1998, a total of 22,828 SMEs went bankrupt, an increase of 33 per cent year on year. The new administration made the promotion of smaller firms a priority in its economic strategy, focusing in particular on the start up of small, focused and flexible high-tech companies. In addition to government funding of W4.0 trillion (US$3.3 billion) a year, a small number of so-called 'Angel Clubs' came into existence, through which wealthy individuals provided funding for start-up ventures. Nevertheless, SMEs continued to face funding problems as banks adopted a more cautious approach to lending and favoured borrowers with collateral to guarantee their loans (http://www.feer.com/9903_04 *Far Eastern Economic Review* 4 March 1999).

Despite the public avowals of commitment to reform by the *chaebŏl*, the actual implementation of restructuring measures has turned out to be a tortuous process. The reaction of some *chaebŏl* leaders to the reform programme was openly critical, with Daewoo chairman Kim Woo-choong publicly stating that the reforms had been formulated by the IMF with the aim of rendering Korean corporations unable to compete in global markets. In other words, the reforms were an attempt by the United States and Japan to tame the *chaebŏl* (Kim Chong-tae 1998i: 13 and 1998o: 34–35).

Writing in early 1998, Kim Chong-tae highlighted the struggle that lay ahead for the IMF and the Korean government in their efforts to restructure the *chaebŏl*:

> (The) chaebols are reluctant to restructure their businesses. They would just like to cope with the so-called 'IMF wave' through layoffs and reductions of overhead (sic), instead of the fundamental reform of their business structures (Kim Chong-tae 1998o: 35).

An article in *Business Week* took a similar line, observing that the conglomerates were more likely to keep a low profile, give the appearance of restructuring, and bide their time until an upswing in the economy allowed them to return to their old ways (*Business Korea* February 1998: 21).

By the end of 1998, the Korean press was reporting that the top five were still lagging behind other conglomerates in their restructuring activity, had failed to reduce their debt levels, and were continuing to borrow money and expand their businesses (Kim Citi 1998a: 20–21). An article in the *Economist* noted that the top five had raised W3.2 trillion (US$2.7 billion) in the capital markets in the first ten months of 1998, an increase of more than 400 per cent year on year, and had accounted for 75 per cent of all new bond issues in the same period (http://www.economist.com/archive/ *Economist* 14 November 1998).

In April 1999, there were reports that, while the 'Big Deals' had failed to materialize, the conglomerates were engaging in fierce competition to take over state-run firms that were due to be privatized (*Business Korea* April 1999: 43). Three months later, amidst continuing accusations from the public and media that the conglomerates were dragging their feet, the government began to threaten punitive action against business groups that failed to take an active part in the restructuring and reform process. In early August, the top five agreed with the government to abolish their traditional fleet-style management, under which all affiliates were centrally controlled, and allow the development of independent, world-class companies. On August 15, 1999, President Kim announced to the nation:

> I am determined to go down in Korea's history as a president who first accomplished corporate reforms and straightened things out in the economy. I will remain firm (http://www.feer.com/9908_26 *Far Eastern Economic Review* 26 August 1999).

The following day, the unthinkable happened: Daewoo's creditor banks announced that the Daewoo Group was to be dismantled and published plans for restructuring the conglomerate, which now had the dubious honour of being the single biggest corporate bankruptcy in history. Daewoo, which had been seen as the conglomerate most resistant to reform (McGonagle 2000: 44), had lost US$458 million on sales of US$51 billion in 1998, the year in which it increased its affiliates from 275 to 289 and boosted its debt by more than 40 per cent (http://www.economist.com/archive/ *Economist* 21 August 1999). By the summer of 1999, Daewoo's domestic debt had reached US$50 billion, almost as large a sum as the bailout funds allocated to Korea by the IMF (*Business Korea* September 1999: 23). On 19 July 1999, the management of Daewoo had been forced to seek assistance from its creditor banks in the form of a restructuring of its mounting debts, but on this occasion there was no help to be had.

The Daewoo Group was to be trimmed down to just six companies,[7] with viable affiliates including Daewoo Electronics and Daewoo Securities to be sold off, and others including the shipbuilding and construction divisions to be restructured as independent companies before being sold to foreign or domestic buyers (http://www.feer.com/9908_26 *Far Eastern Economic Review* 26 August 1999). For its part, the government, which was believed to have orchestrated the restructuring plans, agreed to protect Daewoo from being forced into liquidation (http://www.economist.com/archive/ *Economist* 21 August 1999). Although early reports suggested that Chairman Kim Woo-choong would step down, it was later rumoured that he would remain involved in the running of Daewoo Motor, with the shareholders' approval (http://www.feer.com/9908_26 *Far Eastern Economic Review* 26 August 1999).

A year later, it was rumoured that the Hyundai Group would be the next to fall. Having taken over Kia Motors, Hanwha's oil refining division, and other ailing compnaies, Hyundai's debts had risen to US$46 billion and its debt-to-equity ratio had jumped to 784 per cent in 1998 (McGonagle 2000: 43–44). Despite attempts to improve the group's financial standing in 1999, Hyundai came under increasing pressure from the government and its creditor banks to reform and restructure. In June 2000, Chairman Chung Ju-yung announced that he and his two sons would retire from the management of the group, leaving its operations in the hands of professional managers. This was seen by many as an important step towards the breaking down of the traditional *chaebŏl* management structure. Some, however, remained sceptical of Chung's motives and long-term strategy (*Business Korea* June 2000: 12) as it seemed unlikely that the elderly tycoon would be able to resist the temptation to retain some element of control over the empire he had created.

Negative press reports notwithstanding, some progress had been made in reforming and restructuring the *chaebŏl*. Average debt-to-equity ratios had fallen from more than 500 per cent at the end of 1997 to 302 per cent in June 1999, with expectations that many of the conglomerates would meet the government's target of 200 per cent by the end of the year. The leading conglomerates had raised substantial amounts of funds through the induction of foreign capital, sales of non-core affiliates, and equity and bond issues in the capital markets (*Business Korea* October 1999: 23).

As of the summer of 2000, it was still difficult to predict the full impact of these changes in the corporate sector on Korea's outward foreign direct investment. As far as the electronics sector is concerned, news has been scant and, generally speaking, the effects of the crisis have had to be inferred from general reports on developments within the business groups. In early 1998, as the *chaebŏl* announced their restructuring plans, it was reported that Hyundai was planning to pull out of its proposed semiconductor investment in Scotland, and that LG would close down 90 affiliates worth US$1.6 billion by the end of 1999. Samsung was said to be reviewing its investment in the north-east of England and considering a reduction of up to 40 per cent in its investments in Europe. The review was thought to include the transfer of some production facilities from western to eastern and central Europe (Kim Chong-tae 1998m: 16–18). Following the collapse of the Daewoo group in August 1999, it was announced that the electronics division was to be sold to an investment group of American banks for the sum of US$3.2 billion, leaving the future of Daewoo's global sales and production network in the balance for the time being (http://www.economist.com/archive/ *Economist* 21 August 1999).[8]

THE PUBLIC SECTOR

Having identified collusion between the government, big business, and the banks as a factor contributing to the crisis, the new government took steps to bring the nation's regulatory framework up to levels acceptable to industrialized nations and global organizations such as the OECD and WTO. In February 1998, the government established a Regulatory Reform Committee and entrusted it with the task of reducing and simplifying the country's economic and social regulations (Kwon 2000: 38–39). The government also announced its intention to privatize a number of major state-run enterprises, including Pohang Iron and Steel (POSCO), Korea Telecom, the Korea Electric Power Company (KEPCO), and the Korea Gas Corporation (*Business Korea* September 1998: 43).

Government reform measures resulted in a reduction in the number of civil servants, with more than 9,000 employees in the central government and more than 35,000 in local government losing their jobs during the restructuring process. In the first year of its existence, the Regulatory Reform Committee removed 5,430 regulations from the statute books, abolished 23 laws and revised almost 300 others (Kwon 2000: 38–39).

THE LABOUR MARKET

The principal aim of reforms in this area was to improve the flexibility of the labour force, which would facilitate corporate restructuring and enhance the attraction of Korea as a host for inward investment. The establishment of the Tripartite Commission, comprising representatives from government, labour, and management, enabled agreement to be reached on passing the necessary legislation to achieve this goal. The new laws provided for layoffs in times of need (such as in the wake of a hostile takeover) and thus brought to an end the expectation of lifetime employment among Korean workers (Kwon 2000: 37–38). Given that restructuring would inevitably lead to higher levels of unemployment, the government made a commitment to devote more resources to expanding the social safety net by extending the employment insurance scheme to all workers and increasing funding for unemployment protection measures (*Korea Trade Update* November 1999: 3).

Between 1996 and 1998, almost two thirds of Korean companies, including more than 70 per cent of the country's large firms, implemented downsizing policies, reduced wages for the first time in Korea's history, and began to increase the proportion of part-time employees in their workforce. (The proportion of part-timers rose from 18.8 per cent in 1996 to 42.3 per cent in 1998) (Kwon 2000: 38). For its part, the government increased spending on unemployment protection measures from US$4.7 billion in 1998 to US$14.0 billion in 1999 (*Korea Trade Update* November 1999: 3). The action was timely; in the first three months of 1998 more than 700,000 Koreans lost their jobs, bringing the unemployment rate up to a 12–year high of 6.5 per cent, with 1.3 million unemployed (*Business Korea* May 1998: 11). However, worse was to come; by February 1999, the jobless rate had soared to 8.6 per cent, with 1.8 million Koreans looking for work (http://www.feer.com/9911_04 *Far Eastern Economic Review* 4 November 1999).

The unemployment figures could have been much worse, had the government not encouraged companies to limit layoffs as far as was possible. The Korean and western press reported cases of companies putting workers on unpaid 'paternity leave', and Hyundai's plans to let 5,000 workers go at its Ulsan plant resulted in actual layoffs of 277 employees, with the rest due to return to work after 18 months' unpaid leave (http://www.economist.com/archive/ *Economist* 10 July 1999). A turning point in labour relations came in April 1999, when striking subway workers were given an ultimatum: return to work or join the ranks of the unemployed. Their decision to end the strike prompted the *Far Eastern Economic Review* to observe that, with lifetime employment a thing of the past, unions no longer had the luxury of taking action safe in the knowledge that jobs would not be lost (http://www.feer.com/9911_04 *Far Eastern Economic Review* 4 November 1999).

TRADE AND FOREIGN DIRECT INVESTMENT (FDI)

The major reforms implemented in the area of trade were the liberalization of capital account transactions and the abolition of restrictions on trade in general, and on imports from Japan in particular (Suh 1998: 36 and Kim In-June 1998: 7). The Korean press noted that the Japanese had identified restrictions on imports as a key factor hindering the development of FDI in Korea. In 1998, the Import Source Diversification Programme was abolished, eliminating restrictions on the imports of 16 items from Japan, including machinery and electronics products. (The programme had been implemented in 1981 with the dual aims of reducing Korea's trade deficit and protecting domestic industry [*Business Korea*, May 1998: 34].) By the end of 1999, the domestic market had been opened to all foreign goods and services (with the exception of rice and beef); the system of customs approval had been changed from pre-approval to notification after the fact, and tariffs on manufactured goods had been reduced to a trade-weighted average of 6.2 per cent (*Korea Trade Update* November 1999: 4 and Kim June-dong 2000: 39).

Perhaps one of the most significant areas of reform from the point of view of foreign companies was that of inward foreign direct investment. While the need for Korean companies to globalize their operations through outward FDI had been recognized and promoted since the late 1980s, the Koreans had been less keen to act as hosts to inward investment, unless it brought with it necessary technology and know-how. However, President Kim Dae-jung made it clear in the early days of his administration that the old attitudes toward inward FDI would have to change:

> We have been hostile toward foreign capital and have a negative national sentiment towards (foreign investors). But we have to treat them as valuable (sic) guests (*Business Korea* January 1998: 18–19).

Foreign direct investment was now seen as a means of bringing much-needed capital, know-how and competitiveness to Korean industry, and, for the *chaebŏl*, an opportunity to raise capital through the sale of assets and equity, in the absence of any major buyers or investors in Korea (Hur 1998b: 22–23). Restrictions on equity participation in Korean companies were lifted, and a law was passed to allow mergers and acquisitions of domestic firms. As Kim Chong-tae observed in 1999:

> Before the crisis, the transfer of management rights of Korean companies to foreign firms was unthinkable. Now, domestic firms are hurriedly selling their assets to introduce foreign capital as a way to restore corporate health (Kim Chong-tae 1999a: 27).

Measures were taken by the government with the aim of creating a favourable investment environment that would propel Korea to the ranks of the top ten host nations for FDI by 2002. The Foreign Investment Promotion Act, which went into effect in November 1998, aimed to improve the support systems for foreign investors. More than half the regulations relating to FDI were abolished between 1998 and 1999, and the Korea Investment Service Centre was established to provide a 'one-stop shop' for potential investors seeking information about the market, investment opportunities, and the regulatory environment. Regional governments also moved quickly to set up information and advisory services, in an effort to attract inward FDI to the provinces. In 1999, the government appointed an ombudsman to deal with issues raised and problems faced by foreign investors in Korea, and assigned 30 'home doctors' to visit foreign businessmen in their offices to deal with comments and complaints (Kim June-dong 2000: 39 and McGonagle 2000: 42).

In 1998, a total of 22 sectors were opened up to inward investment (including real estate rentals and sales, and land development), with a further five being liberalized in May 1999 (book publishing, alcoholic beverages, and casinos being among them). These measures left just 21 sectors in which foreign investment is partly or wholly restricted, mainly areas with links to national security, agriculture, and the protection of national culture (Kim June-dong 2000: 39).

As part of the reform package, restrictions on the purchase of Korean real estate by foreigners and foreign companies were lifted. President Kim went so far as to include a team of real estate experts in the investment promotion mission that accompanied him on his visit to the United States in 1998. From July of that year, foreigners were allowed to buy office buildings, industrial sites, apartments and land. It was estimated that, in total, US$35 billion in corporate real estate was being offered for purchase by foreign buyers (Kim Chong-tae 1998g: 20–24).

The impact of these reforms on inward FDI was dramatic: annual investment, which had fluctuated between US$1 billion and US$3 billion between 1993 and 1996, rose to US$7.0 billion in 1997, before soaring to new record levels of US$8.9 billion in 1998 and US$15.5 billion in 1999 (*Korea Trade Update* November 1999: 4). A notable feature of the 1998 figure was that mergers and acquisitions activity accounted for almost 40 per cent of the total (US$3.5 billion). Major investors were the United States (US$2.97 billion) and the European Union (US$2.89 billion) (Kim Chong-tae 1999a: 26). However, M&A activity slowed in 1999, as interest in joint ventures began to accelerate (Kim June-dong 2000: 36–37). As of the end of January 1999, foreign investors owned a majority share in 42 listed companies in Korea and, by the end of March 1999, the top five *chaebŏl* had attracted US$3.9 trillion in foreign capital (Kwon 2000: 36).

THE KOREAN ECONOMY SINCE THE 1997 CRISIS

In December 1997, the IMF forecast a GDP growth rate of just 1 per cent for Korea in 1998. This was lower than the Korean government's forecast of 3 per cent, and some private research institutes in Seoul were warning that the economy might even experience negative growth, for the first time since 1980 (*Business Korea* January 1998: 36 and Kim In-June 1998: 9). The more pessimistic analysts were correct, as the Korean economy shrank by 5.8 per cent during its first year under the IMF programme, with per capita GNP falling back to levels seen in the early 1990s (US$6,823). In 1998, South Korea ranked 17th in the world economy, down from 11th place the year before (Sohn Young-ju 1999a: 19).

The poor performance in 1998 reflected a decline in private consumption, as unemployment levels rose and concerns grew over possible job layoffs, and slower export performance as the yen weakened and key export markets in Asia remained mired in recession. Prior to the crisis, Asian countries had accounted for half of South Korea's export sales; in 1998, exports to the ASEAN nations fell by 30 per cent in value terms (http://www.economist.com/archive/ *Economist* 12 September 1998 and *Business Korea* April 1999: 22). In the first half of the year, facility investment plunged 48.6 per cent year on year, while machinery and domestic construction orders were down by 47.0 per cent and 58.6 per cent respectively (*Business Korea* July 1998: 13). At 63.7 per cent, the factory operating ratio was at its lowest level since records began in 1985 (Kim Chong-tae 1998c: 14).

A key element of the IMF's economic policy was the maintenance of a tight monetary policy and high interest rates to encourage saving and induce foreign capital into the nation's capital markets. However, the inflow of foreign funds was limited due to concerns over the prospects for the Korean economy, and the continuing credit crunch and high interest rates aggravated the problems facing domestic firms in terms of raising capital and servicing their debts. At one point in 1998, more than 3,000 businesses were going bankrupt every month (http://www.economist.com/archive/ *Economist* 31 July 1999).

With the rate of economic growth plunging to minus 5.3 per cent in the first half of the year, the IMF and Korean government announced a revision of their economic policy in July 1998. With usable foreign exchange reserves expected to reach US$43 billion by the end of the year (exceeding the IMF's target of US$41 billion) and thus improve the country's external liquidity position, focus in the second half of the year was to be placed on promoting a recovery in domestic demand, while pushing ahead with reform in the corporate and financial sectors. The new policy called for an easing of monetary policy, the lowering of interest rates, an increase in spending on the social safety net and social overhead capital to assist the unemployed and create new jobs, and a reduction in special consumption tax to encourage consumer spending (*Business Korea* August 1998: 20–21).

Despite this action, private consumption remained stagnant and economic activity failed to pick up, prompting the government to announce a further set of measures to boost the economy in November 1998. Additional funding would be made available for export companies, SMEs and firms engaged in domestic construction, and consumers would have access to more credit for purchases of consumer durables. A mortgage system was to be introduced to encourage house buying, and a series of infrastructure projects would be promoted in order to boost the domestic construction industry (Kim Citi 1998b: 12–13).

By the end of the year, exchange rates had stabilized with the wŏn/dollar exchange rate recovering from W1,900/$ to W1,200/$ during the year, interest rates had fallen to less than 5 per cent (from more than 12 per cent in 1996), usable foreign exchange reserves had reached US$48 billion, and the trade and current accounts had accumulated record surpluses of US$40.5 billion and US$39.0 billion respectively (*ROK Economic Bulletin* December 1999: 2–3, Kim In-June 1998: 14–15 and Kim Chong-tae 1999d: 15–16).

However, analysts noted that the improvement in the balance of payments situation was due to a fall in imports rather than a recovery in export performance. Furthermore, facility investment had fallen by 38.5 per cent year on year, and the unemployment rate had soared from 2.1 per cent in October 1997 to 7.9 per cent in December 1998 (Sohn Young-ju 1999a: 19 and Kim Chong-tae 1999d: 16). Concerns remained over the potential impact of developments in the global economy, including the performance of the US economy, increases in the prices of oil and other essential raw materials, the potential devaluation of the Chinese currency, and the pace of recovery in Asian export markets.

Nevertheless, towards the end of 1998, rising levels of inward foreign direct investment and a surge in foreign participation in the Korean stock market boosted confidence in the economy, resulting in a recovery in domestic demand. Private consumption began to show signs of picking up as the stock market and real estate market regained their lost momentum.

In the first half of 1999, South Korea recorded a GDP growth rate of 7.3 per cent, placing it first among OECD member nations in terms of economic performance. The Republic also ranked third in terms of usable foreign exchange reserves, which had increased to US$62.0 billion (*ROK Economic Bulletin* December 1999: 2–3). In May, exports recorded a positive growth rate after a four-month decline, with the year-on-year growth rate rising to 18.6 per cent by July, as export performance was boosted by strong demand from industrialized countries and the ASEAN nations (*Business Korea* July 1999: 12–14 and September 1999: 14–16). The first half of the year saw imports of capital and consumer goods increase by 25.5 per cent and 15.7 per cent respectively, and the factory operating ratio, which had hovered around the 60 per cent mark in 1998, recovered to more than 80 per cent in July (*Business Korea* September 1999: 14– 16 and October 1999: 22). In response to these favourable indications of economic recovery, the government once again revised its economic policy, this time with the aim of maintaining the *status quo*, rather than continuing to boost the economy.

Although key economic indicators were undoubtedly improving, concerns remained that the recovery would be 'W'- rather than 'V'-shaped, with Korea plunging back into recession after a short-lived recovery. Analysts expressed unease that the apparent improvement in economic performance had more to do with the depths to which the economy had sunk in 1998 rather than the heights to which it could climb on the basis of achievements in economic and corporate restructuring (Sohn Cecile 1999: 16–17).

In his speech to the nation on 3 December 1999 (the second anniversary of the signing of the IMF agreement), President Kim noted that, as promised, it had taken just 18 months for the Korean economy to recover to pre-crisis levels. Rescue loans extended by the IMF had been repaid, usable foreign exchange reserves stood at record levels, interest and exchange rates were stable, and Korea's credit ratings had been restored to investment grade (*ROK Economic Bulletin* December 1999: 57).

In October 1997, as concerns mounted over the crisis in the East Asian markets in general and the soundness of the Korean economy and financial markets in particular, Moody's Investor Service had downgraded the credit rating of the Korea Exchange Bank and four other leading banks. The move reflected concerns over the banks' heavy lending to the conglomerates and the poor quality of their asset portfolios (Kim Chong-tae 1997b: 17). A month later, Moody's lowered Korea's sovereign rating from A3 to Ba1, effectively reducing the nation's long-term bonds from investment grade to junk bond status. This had added to the feeling of nervousness circulating among foreign lenders and investors and made it even harder for the Koreans to obtain financing in international markets (*Business Korea* January 1998: 55).

A year later, Moody's had remained pessimistic about the prospects for Korean banks, expressing particular concern over the inadequacy of their capital and their thin profit margins, and warning that further downgrades might follow in 1999 (Kim Citi 1998c: 27–28). However, in early 1999, Moody's, Standard and Poor's and Fitch IBCA all raised Korea's sovereign rating back to investment grade, citing the improved levels of foreign exchange reserves, reduction in short-term debt, and achievements in the reform of the financial and corporate sectors (*Business Korea* March 1999: 10–11). In June, Fitch IBCA raised Korea's foreign currency rating from BBB- to BBB, citing the 'strong reassertion of the nation's traditional economic fundamentals' (*Business Korea* July 1999: 74).

In his anniversary address to the nation, President Kim also struck a cautionary note, warning against the dangers of complacency:

> Two years ago today was regarded as the most shameful day for the Korean people. Two years later, we mark today not as a proud day for us to proclaim the end of the foreign exchange crisis but rather as the day to renew our resolve. We must not forget why the crisis, which brought pain and trials, occurred (*ROK Economic Bulletin* December 1999: 56).

In 1999, a robust recovery in private consumption, exports and facility investment fuelled a GDP growth rate of 10.7 per cent; a remarkable achievement, given that the IMF's forecast at the beginning of the year had been for 2 per cent growth. Private consumption had risen to reach 140 per cent of pre-crisis levels, boosted by rising income and gains from the real estate and stock markets. In 1999, the Korea Stock Exchange's market capitalization soared 148 per cent to reach US$285 billion, with the stock price index recovering to pre-crisis levels (*Business Korea* February 2000: 35–36). Exports grew by more than 17 per cent year on year, as global demand increased and the yen strengthened against the dollar, enhancing the relative price competitiveness of Korean exports (*Business Korea* January 2000: 14).

However, the recovery brought new concerns that, as the economy regained its vigour, the will to reform might begin to diminish. With unemployment levels falling back to less than 5 per cent by the autumn and average wages rising by 8 per cent year-on-year, labour unions once again began to flex their muscles, trying to regain the wage levels and benefits that had been forfeited during the recession (http://www.feer.com/9911_04 *Far Eastern Economic Review* 4 November 1999).

An article written by a researcher at the Samsung Economic Research Institute and published in the *Korea Herald* in January 2000 identified seven problems facing the Korean economy at the start of the 21st century. The problems were the possibility of a hard landing for the US economy, the continuing instability in the Korean financial markets, the polarization of Korea's industrial structure into booming information technology-related industries and 'fading' traditional manufacturing industries, and the widening income gap in Korean society. Added to these were the problems of the growing government budget deficit, the re-appearance of inflationary pressures (especially ahead of the April elections), and the possibility of bubbles appearing in the stock market, leaving it vulnerable to changes in the global and domestic economies (http://www.koreaherald.co.kr/ *Korea Herald* 17 January 2000).

Other potential problems identified by analysts included the volatility of international oil prices, the impact on Korean exports of a weakening of the yen and a strengthening of the wŏn, wage increases outpacing productivity gains, and surging imports of consumer and capital goods as private consumption and facility investment gathered momentum. Indeed, the trade account, which had shown a surplus for 26 consecutive months from November 1997, fell back into the red in January 2000, as Korea's export competitiveness was once again threatened by currency fluctuations and increases in production factor costs (*Business Korea* March 2000: 12–13).

In late June, the Korean government unveiled its economic management plan for the second half of 2000, calling for GDP growth of 8 per cent, a jobless rate of 4 per cent, a current account surplus of US$10–US$12 billion, and inflation of 2.5 per cent or less at the consumer level. The principal goals for economic management were announced as the expansion of foundations for sustainable economic growth, the completion of the second phase of reform in which the principles of transparency, accountability, and a free market economy would become deeply entrenched in Korea, and 'expediting the transition towards a digital and knowledge-based economy' (http://www.koreaherald.co.kr/ *Korea Herald* 24 June 2000 and *Business Korea* February 2000: 41).

The government's forecast for GDP growth in 2000, which had been raised from the original projection of 6 per cent following strong growth in the first half of the year, was lower than the IMF and OECD's forecast of 8.5 per cent expansion. With exports forecast to rise by 18.0 per cent to US$170 billion, and imports expected to expand by more than 30 per cent to reach US$160 billion, the trade surplus was forecast to reach around US$10 billion, down from US$24 billion in 1999 (http://www.koreaherald.co.kr/ 24 June 2000).

By the summer of 2000, the Korean media was expressing concern over the prospects for a continued recovery, as the trade surplus shrank, the stock market fell, and reform progressed at 'a snail's pace'. At US$770 million, the trade surplus for the first four months of the year was just 10 per cent of the surplus accumulated in the whole of 1999, threatening the levels of foreign exchange reserves needed to finance the nation's swelling import bill. In the first four months of 2000, imports had soared by 51 per cent year on year, compared with a growth rate of 27 per cent for exports (*Business Korea* June 2000: 23). The stock market index had fallen back from 1,028 at the end of 1999 to 725 by April 2000, with an even sharper fall in the technology-heavy KOSDAQ index, from 256 to 128.

Despite these trends, government officials remained optimistic about the economy, brushing off speculation about the possibility of a second crisis and citing low inflation, rising investor confidence (as evidenced by record levels of inward FDI activity), and the US$84.6 billion now held in foreign exchange reserves (*Business Korea* June 2000: 18–19). Officials also drew attention to the fact that Korea had become a creditor nation in 2000, with US$146.5 billion in external credit and US$136.8 billion in foreign debt (*Business Korea* April 2000: 14–15).

Nevertheless, the same concerns continued to nag at investors and analysts, including the apparent weakening of resolve to carry through reforms to the bitter end. Another worry was the continuing weakness of the financial sector, into which the ROK government had already pumped US$79.6 billion, but which still faced the nightmare of bad loans escalating even further should restructuring bring more bankruptcies. There were signs of an economic recovery, but the long-term prospects for growth were dependent on finishing the job that had been started in December 1997. The situation facing Korea in mid-2000 was stated succinctly in a report published by the OECD:

> Korea is at the moment walking a tightrope – it can neither stop nor turn round without the risk of falling (*Business Korea* June 2000: 20).

In the summer of 2000, prospects for Korean FDI in the 21st century were uncertain. While the rationale for the creation of overseas production and sales networks remains compelling, future investments will have to be made on a more secure financial footing than in the past. The *chaebŏls'* rapid and often reckless expansion has been identified by Korean and western analysts as a factor in Korea's financial and economic problems in the late 1990s, and the suspension or withdrawal of major FDI projects overseas after the 1997 crisis may give host countries and governments food for thought when considering future investment proposals from Korean multinationals.

However, provided that the current programme of reform and restructuring is carried through to a successful conclusion, the *chaebŏl* should, in the medium to long term, be able to adopt a more rational and focused approach to internationalization, establishing a global presence as world-class, financially sound corporations. The links forged with Europe over the past decade will surely grow stronger, given the mutual recognition of the benefits of such a partnership and the interest demonstrated in creating new channels of discussion and cooperation. The crisis of 1997 has created a new perception of the EU nations as 'all-weather friends', with a shared interest in boosting bilateral trade and investment in the 21st century.

Notes

INTRODUCTION

1 Throughout this book, the Republic of Korea (ROK) is referred to as 'Korea', 'South Korea', or 'the Republic'. The Democratic People's Republic of Korea (DPRK) is referred to as 'North Korea'.
2 The Korean government statistics used in this thesis classify the following countries as 'Europe': Austria, Belgium, Czech Republic, France, Germany, Hungary, Ireland, Italy, Kazakhstan, Malta, Netherlands, Poland, Portugal, Romania, Russian Federation, Slovakia, Slovenia, Spain, Switzerland, Tadzhikistan, Turkey, United Kingdom, and Uzbekistan. (See Chapter 5)
3 The term 'upstream foreign direct investment' has also appeared in the literature, for example, in Jun and Prendergast (1991).
4 Although many Korean writers have made contributions to the debate on FDI theory, with the exception of Byun Jae-woong's work published in 1995, these have not appeared in English, and none of the works written in English by Korean scholars has made reference to Korean contributions to the debate.

CHAPTER 1

1 Summarized in Dunning 1993b: 74.
2 In the 1970s and 1980s the debate on who first applied the concept of internalization to the theory of FDI focused on the works of Williamson (1975) and Buckley and Casson (1976). However, Buckley himself acknowledged that Stephen Hymer was the pioneer in this field (Buckley 1992: 4); a paper published in French by Hymer in 1968 predated the revival in interest in Coase's work and transaction cost economics by almost a decade.

CHAPTER 2

1 In 1991, Jun and Yun co-authored a paper entitled 'An explanation for Korea's foreign direct investment in developed countries: Structural market defence approach'.
2 A fourth thesis, written by Hong Sunghoon, was submitted in 1995 but the University of Warwick declined to lend the work to the author.

CHAPTER 3

1 The crisis that swept through Korea in 1997 was the result of a broad range of domestic and external factors. Although it is beyond the scope of this work to consider these factors in detail, such an analysis will form the foundation of the author's next research project (*'Korean FDI after the 1997 crisis'*) to be undertaken from September 2000.

2 The definition of an SME is a firm with fewer than 300 full-time employees in the mining, manufacturing and transportation sectors, fewer than 200 in construction, and fewer than 20 in the trading and service sectors (Cho 1994: 64).

CHAPTER 4

1 The author was unable to obtain a copy of this thesis from the University of Warwick.

2 In the wake of the economic crisis in 1997 and 1998, the top five *chaebŏl* came under pressure to restructure; one of the 'Big Deals' proposed as a means of achieving this aim was the takeover of Daewoo Electronics by Samsung Electronics. However, in August 1999, the Daewoo Group was dismantled, and it was announced that the electronics division was to be sold. A year later, there were reports that the proposed sale to a US investor had fallen through.

CHAPTER 5

1 FDI statistics are quoted throughout on a net invested basis: that is, investment net of liquidations and withdrawals.

2 The AC markets are: Japan, France, Italy, Germany, the Netherlands, Belgium, Norway, Sweden, Finland, the United Kingdom, Switzerland, Austria, Portugal, Eire, Spain, Iceland, the United States, Canada, Australia, and New Zealand.

CHAPTER 6

1 Throughout this book, the definition of 'Europe' used in presenting official Korean FDI statistics has been used, as described in the introduction. Extending the analysis to the countries of central Europe (including Uzbekistan and Khazakstan) enables us to include in the study a wide range of countries with vastly different advantages to offer to Korean investors.

2 In this chapter, the term 'Korea' is used as the name of the unified country that existed until 1948. Thereafter, the Republic of Korea (ROK) is referred to as 'Korea', 'South Korea', or 'the Republic'. The Democratic People's Republic of Korea (DPRK) is referred to as 'North Korea'.

3 The three exceptions were Portugal, Sweden and Denmark.

4 The 25 founder nations of ASEM are the 15 members of the EU (Austria, Belgium, Denmark, Finland, France, Germany, Greece, Ireland, Italy, Luxembourg, the Netherlands, Portugal, Spain, Sweden, and the United Kingdom), and Brunei, Indonesia, Japan, Malaysia, the People's Republic of China, the Philippines, Singapore, South Korea, Thailand, and Vietnam. The President of European Commission is also a participant at ASEM.

5 Subsequent meetings were held in London (April 1998) and Seoul (October 2000), with the 2002 meeting due to be hosted by one of the European nations.

6 The three main initiatives arising from ASEM 1 were the Asia-Europe Foundation (a forum for cultural and humanitarian issues), the Asia-Europe Business Forum (a private sector-led initiative focusing on trade, investment, and business collaboration), and the Investment Promotion Action Plan (an initiative to develop a framework for the promotion and regulation of multilateral investment) (Windham-Wright 1999: 21-22).
7 The Korean term used by Kim Bak-soo is *Pŏm yurŏp kyŏngjegwŏn*, whereas Ŏm Ch'ang-sŏp uses *Yurŏp kwangnyŏk kyŏngjegwŏn*.

CHAPTER 8

1 Early investments in Europe by the company now known as LG Electronics were made under the name of Goldstar; for the sake of clarity, the name 'LG (Electronics)' will be used to refer to the company throughout this section.
2 The 'Big Deals' included a proposal to swap Daewoo's electronics division for Samsung's automobile division, thus removing Daewoo from the consumer electronics sector. However, the proposed swap did not materialize, and Daewoo Electronics was put up for sale after the Daewoo Group was dismantled in the summer of 1999.
3 A full discussion of relations between the United States, North and South Korea, and the nuclear crisis of 1993-1994 is beyond the scope of this book. Detailed analysis of these and other related issues can be found in the monographs listed in the bibliography, including Scalapino and Lee 1986, Sutter and Han (1990), and Kihl and Hayes (1997).

SUMMARY AND CONCLUSIONS

1 The data published by the Ministry of Finance and Economy was on a total permitted basis; the data used in this study is on a net permitted basis. The net permitted amount for 1996 was US$3.5 billion.

POSTSCRIPT: THE *CHAEBŎL* AND THE KOREAN ECONOMY IN THE 'IMF ERA'

1 The FSC has two subsidiaries: the Financial Supervisory Service (FSS) and the Securities and Futures Commission.
2 An exchange rate of W1,200/US$ is used throughout this chapter.
3 Dae Dong was taken over by Kookmin Bank, Dong Nam by the Housing and Commercial Bank, Dong Hwa by Shinhan Bank, Kyungki by KorAm Bank, and Chung Chong by Hana Bank.
4 Although HSBC agreed to pay US$900 million for a 70 per cent stake in Seoul Bank, negotiations foundered in 1999 on the issue of asset valuation, especially with regard to non-performing loans (http://www.economist.com/archive/ *Economist* 4 September 1999).
5 The full list of industries is: automobiles, ships, steel, aircraft, computers, petrochemicals, power generation equipment, rolling stock, LCD displays, and braun tubes.
6 In early 1999, discussions were still unresolved; Daewoo and Samsung had failed to conclude their negotiations, while Hyundai and LG could not reach an agreement on asset valuation (Kim Chong-tae 1999d: 16-17). The demise of Daewoo in the summer of 1999 put an end to hopes for one of these two 'Big Deals'.

7 Daewoo Motor, Daewoo Motor Sales, Daewoo Corp. (minus the construction division), Daewoo Heavy Industries (minus the shipbuilding division), Daewoo Telecom, and Daewoo Capital.

8 In July 2000, there were unconfirmed reports that the sale of Daewoo Electronics had fallen through.

Appendices

Appendix A Structure of Global Electronics Production: By Sector, 1996

	World	*USA*	*Japan*	*Germany*	*UK*	*Singapore*	*Taiwan*	*ROK*
Output US$bn	10,332	2,985	2,824	524	375	427	317	506
Ind. Electronics %	63	71	55	71	74	62	68	28
Pers. Computers	27	27	29	16	34	56	58	16
Telecommunications	19	27	17	24	23	4	8	10
Other	16	17	9	31	17	2	2	2
Consumer Electronics %	8	2	8	6	7	7	4	18
Parts & Components %	29	17	27	23	19	31	28	54
Semiconductors, Colour picture tubes	18	15	23	9	8	21	10	40
Other	11	2	4	14	11	10	18	14

Source: EIAK 1997a: 2.

Appendix B Republic of Korea: Electronics Production, 1975–1985

	1975	*1976*	*1977*	*1978*	*1979*	*1980*	*1981*	*1982*	*1983*	*1984*	*1985*
Industrial Electronics US$mm	35	56	103	103	163	169	173	229	298	687	905
% Total	5.7	5.4	9.3	7.6	9.1	8.6	8.0	11.1	11.0	15.4	19.7
Consumer Electronics US$mm	199	390	436	654	925	1,020	1,198	926	1,164	1,794	1,860
% Total	32.2	37.6	39.4	48.1	51.9	51.9	55.2	44.8	43.0	40.3	40.5
Parts & Components US$mm	384	591	568	602	695	775	799	914	1,242	1,970	1,825
% Total	62.1	57.0	51.3	44.3	39.0	39.5	36.8	44.2	45.9	44.3	39.8
Total US$mm	618	1,037	1,107	1,359	1,783	1,964	2,170	2,069	2,704	4,451	4,590

Sources: EIAK 1993a: 5 and EIAK 1993b: 12.

Appendix C Republic of Korea: Electronics Production, 1986–1996

	1986	*1990*	*1991*	*1992*	*1993*	*1994*	*1995*	*1996*
Industrial Electronics US$mm	1,518	6,345	7,104	7,221	8,470	9,887	13,017	14,563
% Total	20.8	21.9	21.5	21.6	22.6	21.0	20.4	23.7
Consumer Electronics US$mm	2,411	10,141	11,054	10,540	11,399	12,621	13,874	13,668
% Total	33.1	35.1	33.4	31.6	30.4	26.8	21.8	22.3
Parts & Components US$mm	3,356	12,432	14,946	15,631	17,621	24,575	36,813	33,136
% Total	46.1	43.0	45.1	46.8	47.0	52.2	57.8	54.0
Total US$mm	7,285	28,918	33,104	33,392	37,490	47,083	63,704	61,367

Source: EIAK 1997b: 16.

Appendix D Republic of Korea: Electronics Exports, 1986–1996

	1986	*1990*	*1991*	*1992*	*1993*	*1994*	*1995*	*1996*
Industrial Electronics US$mm	905	3,481	3,895	4,227	5,230	5,802	7,666	8,400
% Total	19.7	20.2	20.1	20.4	22.0	18.7	17.6	20.4
Consumer Electronics US$mm	1,860	5,727	6,054	5,966	6,461	7,319	7,861	7,836
% Total	40.5	33.3	31.3	28.8	27.2	23.6	18.0	19.0
Parts & Components US$mm	1,825	8,016	9,385	10,490	12,047	17,827	28,065	24,987
% Total	39.8	46.5	48.6	50.8	50.8	57.7	64.4	60.6
Total US$mm	4,590	17,224	19,334	20,683	23,738	30,948	43,592	41,223

Source: EIAK 1997b: 17.

Appendix E Republic of Korea, Major Electronics Products: By Sector, 1970–1990

	1970	*1975*	*1980*	*1985*	*1990*
Consumer Electronics	Radios	Cassette players, black & white televisions etc	Car stereos, colour televisions etc	Video cassette recorders, microwave ovens etc	Camcorders, compact disc players etc
Industrial Electronics	Mechanical converters/ switches	CB transceivers, desktop calculators	Telephones, semi-electronic exchanges	Computer, terminals, cordless phones, photocopiers	Computers, optical electronics, telecom equipment, faxes etc
Parts & Components	Semiconductor assembly	Condensers, transformers	B&W cathode ray tubes, audiotapes etc	Colour cathode ray tubes, television tuners etc	High density integrated circuits, magnetic heads etc

Source: Pae *et al* 1990: 29.

Appendix F Korean and Foreign Electronics Companies: Production and Exports, 1980–1988

	1980		*1985*		*1986*		*1987*		*1988*	
	Prod.	*Exp.*	*Prod.*	*Exp.*	*Prod.*	*Exp.*	*Prod.*	*Exp.*	*Prod.*	*Exp.*
Consumer Electronics										
Korean %	76.0	74.0	88.9	83.4	89.5	85.8	91.2	90.2	92.4	91.7
Foreign %	18.0	23.0	5.6	9.0	4.9	6.9	3.5	4.8	2.9	4.1
JVs %	6.0	3.0	5.4	7.6	5.6	7.3	5.2	5.0	4.7	4.2
Total %	100.0	100.0	100.0	100.0	100.0	100.0	100.0	100.0	100.0	100.0
Value US$bn	1.1	1.0	2.4	1.5	4.8	3.1	7.0	4.9	9.2	6.3
Electronics Industry										
Korean %	48.5	47.6	63.4	66.1	64.3	66.5	65.2	68.9	70.9	73.6
Foreign %	24.3	37.4	13.8	22.3	13.8	20.9	10.5	14.6	8.8	12.2
JVs %	27.1	15.0	22.8	11.7	21.9	12.6	24.3	16.5	20.3	14.2
Total %	100.0	100.0	100.0	100.0	100.0	100.0	100.0	100.0	100.0	100.0
Value US$bn	2.9	2.0	7.3	4.3	12.1	7.4	17.4	10.8	23.5	15.2

Source: Pae *et al* 1990: 37.

Appendix G Korean Electronics Industry Investment in Europe, June 1997.[1]

Mother Firm	*Overseas Company*	*Location*	*Main Products*	*S/P*[2]	*Established*	*Capital*[3]	*SH*[4] *%*
Dae Ryung Precision	Radix (GmbH)	GER	Satellite receivers	S	1994.2	1.4	100
	Radix Telecom Ltd	UK	Satellite receivers	P	1995.4	5.0	100
Daewoo Electro-Components	Daewoo Electro-Components U.K. Ltd	UK	Tuners	P	1995.2	1.0	100
Daewoo Electronics	Daewoo Electronics France S.A.	FRA	MWO	P	1988.10	6.0	100
	Daewoo Electronics U.K. Ltd	UK	VCR	P	1988.11	17.0	100
	Daewoo Electronics S.A.	FRA	CTV, VCR, MWO, MON, AUD	S	1991.3	20.0	100
	Daewoo Electronics Manufacturing S.A.	FRA	CTV	P	1992.4	6.6	100
	Daewoo Electronics España S.A.	SPA	CTV, VCR, MWO, MON	S	1992.4	3.4	100
	Daewoo Electronics Deutschland GmbH	GER	CTV, VCR, MWO, MON, AUD	S	1992.7		
	Daewoo Electronics Hungary KFT	HUN	CTV, VCR, MWO, WM, RFR, AUD	S	1993	1.0	100
	Daewoo Electronics Sales U.K. Ltd	UK	CTV, VCR, MWO, AUD	S	1993.5	3.8	100
	Uz Daewoo Electronics Co.	UZB	CTV, VCR, C/AUD	P	1993.5	5.8	86
	Daewoo Electronics Manufacturing Poland	POL	CTV, WM, deflection yokes, fly back transfers	P	1993.8	13.0	100
	Daewoo Electronics Positron Co.[5]	RUS	CTV, VCR, MWO, WM, RFR	S	1993.9	0.5	100
	Povolzhie Daewoo Co. Ltd	RUS	CTV, VCR, MWO, RFR, WM	S	1993.9	0.5	100
	Daewoo Electronics SPOL S.R.O.	CZR	CTV, VCR, MWO, WM, RFR, AUD	S	1993.10	1.0	100
	Daewol Electronics Poland SP.ZO.O	POL	CTV, VCR, MWO, WM, RFR	S	1993.12	2.0	100
	Daewoo Electronics Benelux	NL	CTV, VCR, MWO, MON	S	1994.1	1.0	100
	Daewoo Almaty Electronics Co.	KHA	CTV, VCR	P	1994.4	5.0	80
	Daewoo Electronics Italia SPA	ITA	CTV, VCR, MWO, AUD	S	1994.6	2.0	100
	Daewoo Electronics Romania S.R.L.	ROM	CE products	S	1995.11	1.0	100
	Daewoo Electronics Vladi Trading Co.	RUS	CE products	S	1996.7	0.5	100

Mother Firm	Overseas Company	Location	Main Products	S/P[2]	Established	Capital [3]	SH[4] %
	Daewoo Electronics Kiev Co.	UKR	CE products	S	1996.8	0.5	100
Daewoo Telecom	Daewoo Telecom Europe	FRA	PC, faxes, telephones, monitors	S	1992.4	7.6	100
	Aloka-Daewoo	UZB	Exchanges, telephones	P	1996.6	6.5	51
	Dniepr-Daewoo	UKR	Exchanges, telephones	P	1996.8	10.0	50
	Daewoo C&C	UZB	PC, faxes, telephones, monitors, copiers	S	1996.9	0.3	94
Haitai Electronics	Sherwood Europe GmbH	GER	Home & car audio products	S	1985.12	11.0	100
	Haitai Electronics (U.K.) Ltd	UK	Audio products	P	1990.6	3.5	100
Hŭngch'ang	HC Electronics GmbH	GER	Satellite receivers	S	1991	300.0	100
Hyundai Electronics	Hyundai Electronics Deutschland GmbH	GER	Semiconductors, PC, MON, LCD	S	1988.11	12.5	100
	Hyundai Electronics U.K. Ltd	UK	Semiconductors (assembly and test)	S	1995.7	11.0	100
	Hyundai Semiconductor Europe Ltd	UK	Semiconductors	P	1997.1	200.0	100
Kolon	Kolon Imperial Graphics plc	UK	Graphic film	P	1991.10	8.3	100
LG Telecommunications	Electromagnetica Goldstar S.R.L.	ROM	Exchanges	P	1994.8	2.7	50
	LG Telecom Corp.	RUS	Exchanges	S	1995.8	0.8	75
LG Electronics	LG Electronics Deutschland GmbH	GER	CTV, VCR, AUD	S	1980.10	23.4	100
	LG Electronics United Kingdom Ltd	UK	CTV, VCR, AUD	S	1986.12	12.1	100
	LG Electronics North of England Ltd	UK	MWO, CTV	P	1988.10	11.7	100
	LG Electronics France S.A.R.L.	FRA	CTV, VCR, AUD	S	1990.12	3.7	100
	LG Electronics Design Tech Ltd	EIR	CTV, VCR, AUD	R&D	1991.3		
	LG Electronics Magyar KFT	HUN	CTV, VCR, AUD	S	1992.10	1.8	100
	LG Electronics Wales	UK	MON, CPT, deflection yokes, fly back transfers etc	P	1996.10	48.0	100
LG Semiconductors	LG Semicon Germany GmbH	GER	Semiconductors	S	1996.8	DM4.5mm	100
Maxon Electronics	Maxon Systems (London) Ltd	UK	Cellular phones,	S	1988.3	0.1	100
	Maxon Europe Ltd	UK	Trunking	S	1990.12	1.8	100
Orion Electric	Daewoo Orion S.A.	FRA	Colour braun tubes	P	1993.9	10.0	50

Mother Firm	*Overseas Company*	*Location*	*Main Products*	*S/P*[2]	*Established*	*Capital*[3]	*SH*[4] *%*
Saehan Media	Saehan Media Ireland Ltd	EIR	Video pancake	P	1987.5	22.5	100
Sambo Computers	Trigem Computers (U.K.) Ltd	UK	Computers	S	1989.10	£1.2mm	55
Samsung Electro-Mechanics	Samsung Portugal Produtos Electro-Mecanicos S.A.	POR	Tuners, deflection yokes, fly back transfers, drums	P	1990.7	7.0	100
Samsung Display Devices	Samsung Elektronische Bauelement GmbH	GER	CPT	P	1993.1	14.0	100
Samsung Corning	Samsung Corning Deutschland GmbH	GER	Braun tube glass	P	1994.3	DM44.6mm	100
Samsung Electronics	Samsung Electronics GmbH	GER	CTV, VCR, MWO, AUD, MON, PC etc	S	1982	57.0	100
	Samsung Electronics Portuguesa S.A.	POR	CTV, VCR, MWO, faxes etc	S	1982.7	5.3	100
	Samsung Electronics (U.K.) Ltd	UK	CTV, VCR, MWO, AUD, MON, PC etc	S	1984	12.0	100
	Samsung Electronics Manufacturing U.K. Ltd	UK	CTV, MWO, MON	P	1987.2	68.0	100
	Samsung Electronics France S.A.	FRA	CTV, VCR, MWO, AUD, MON, PC etc	S	1988.11	12.3	100
	Samsung Electronica Española S.A.	SPA	CTV, VCR	P	1989.1	11.5	100
	Samsung Electronics Hungarian Co. Ltd	HUN	CTV	P	1989.12	26.0	100
	Samsung Electronics Comercial Iberica S.A.	SPA	CTV, VCR, MWO, faxes, cordless phones etc	S	1990.4	17.6	100
	Samsung Electronics Italia S.P.A.	ITA	CTV, VCR, AUD, faxes, cordless phones etc	S	1991.9	15.5	100
	Samsung Electronics Svenska AB	SWE	Faxes, hand phones, CTV, keyphones etc	S	1992.6	4.0	100
	Samsung Electronics Netherland B.V.[6]	NL	MON, PC etc	S	1995.7	1.0	100
	Samsung Electronics Polska	POL	CTV, VCR, MWO, AUD, MON, PC etc	S	1996.2	7.0	100
	Samsung Electronics Overseas B.V.	NL	CTV, VCR, AUD, MWO, MON, PC etc	S	1997.2	0.2	100
SKC	SKC Europe GmbH	GER	Video tape, PET film	S	1987.2	DM1.0mm	100
SKM	SKM Europe Ltd	UK	Audio tape	P	1991.7	2.2	98

Source: EIAK 1997f.

Abbreviations

AUS	Austria	ROM	Romania	AUD	Audio equipment
BLG	Belgium	RUS	Russia	C/AUD	Car audio equipment
CZR	Czech Republic	SPA	Spain	CE	Consumer electronics
DNK	Denmark	SWE	Sweden	CPT	Colour picture tubes
EIR	Eire	TUR	Turkey	CTV	Colour television
FRA	France	UK	United Kingdom	LCD	Liquid crystal display
GER	Germany	UKR	Ukraine	MON	Monitor
HUN	Hungary	UZB	Uzbekistan	MWO	Microwave oven
ITA	Italy			PC	Personal computer
KHA	Khazakstan			PET	Polyethylene terephthalate
NL	Netherlands			RFR	Refrigerator
POL	Poland			VCR	Video cassette recorder
POR	Portugal			WM	Washing machine

Notes

[1] The list is drawn from the EIAK's annual directory of electronics industry overseas investors; the list is limited to members of the EIAK and to those who gave full details of their investment. A supplementary list is given of companies that provided only the most basic information.

[2] Unless otherwise stated, investments are classified as P (production) or S (sales).

[3] Capital invested is given in millions of US dollars, with the exception of LG Semiconductors, Samsung Corning, and SKC (given in millions of Deutsche Marks), and Sambo Computers (given in millions of pounds sterling).

[4] SH = Shareholding in the company. Other abbreviations are given at the end of the table.

[5] Daewoo Electronics has three sales subsidiaries in Russia, located in Moscow (Povolzhie Daewoo), St Petersburg (Daewoo Electronics Positron Co.), and Vladivostok (Daewoo Electronics Vladi Trading Co.).

[6] Samsung Electronics Netherlands B.V. sells Samsung's product in the Netherlands, while Samsung Electronics Overseas B.V. sells to the CIS.

Bibliography

N.B. The bibliography contains all major works that have informed the author's opinion while researching this topic, rather a simple listing of the monographs, articles and statistics cited in the text.

Agmom, Tamir and Kindleberger, C. P. (Ed.) (1977) *Multinationals from small countries*. Cambridge: MIT Press.

Aharoni, Y. (1966) *The foreign investment decision process*. Boston: Harvard University Press.

Ahn Byung-joon. (1998) 'Prospects for Korea under the IMF and Kim Dae-jung', *Korea Focus*, May-Jun, Vol. 6, No. 3, pp. 1–17.

——. (1986) 'North Korea and the United States in a global and regional perspective', in *North Korea in a regional and global context*, edited by Scalapino, Robert A. and Lee Hongkoo. Berkeley: University of California, Institute of East Asian Studies.

Ahn Chung-si. (1998) 'South Korea's relationship with Central Europe and CIS countries: Going beyond the primacy of economy', in *Koreans and Korean business interests in Central Europe and CIS countries*, edited by Ahn Chung-si and Kwon Tai-hwan. Seoul: Seoul National University Press.

Ahn Chung-si and Kwon Tai-hwan. (Ed.) (1998) *Koreans and Korean business interests in Central Europe and CIS countries*. Seoul: Seoul National University Press.

Aliber Robert Z. (1970) 'A theory of direct investment', in *The international corporation*, edited by Charles P. Kindleberger. Cambridge: MIT Press.

Amsden, Alice H. (1989) *Asia's next giant: South Korea and late industrialization*. New York: Oxford University Press.

An Ki-ch'ŏl *et al.* (1991) *EC t'onghap ch'okchin-gwa sanŏp chaep'yŏng panghyang* (The promotion of EC integration and the direction of industrial restructuring). Seoul: Sanŏp Yŏn'guwŏn.

An Se-yŏng. (1995) *Tagukchŏk kiŏp kyŏngjehak* (The economics of the multinational enterprise). Seoul: Pagyŏngsa.

Anderson, Kym and François, Joseph. (1998) 'Commercial links between Western Europe and East Asia: Retrospect and prospects', in *Europe, East Asia and APEC: A shared global agenda?*, edited by Peter Drysdale and David Vines. Cambridge: Cambridge University Press.

Asia-Europe Meeting 1. (1999) 'ASEM 1: Chairman's Statement', in *ASEM: A window of opportunity*, edited by Wim Stokhof and Paul Van der Velde. London: Kegan Paul International.

Bandow, Doug. (1992) 'America's Korean protectorate in a changed world: Time to disengage', in *The US-South Korean Alliance: Time for a change*, edited by Doug Bandow and Ted Galen Carpenter. New Brunswick: Transaction Publishers.

Bandow, Doug and Carpenter, Ted Galen. (Ed.) (1992) *The US-South Korean Alliance: Time for a change*. New Brunswick: Transaction Publishers.

Bark Tae-ho. (1991) *The Korean consumer electronics industry*. Washington: World Bank.

Bridges, Brian. (1999) *Europe and the challenge of the Asia Pacific*. Cheltenham: Edward Elgar.

——. (1988) *Korea and the West*. London: Routledge&Kegan Paul.

British Council (1996) *Korea: A science and technology profile*. London: The British Council.

Buckley, Peter J. (1996) 'Government policy responses to strategic rent-seeking transnational corporations', *Transnational Corporations*, Vol. 5, No. 2, pp. 1–17

——. (1992) *New directions in international business: Research priorities for the 1990s*. Aldershot: Edward Elgar.

——. (1991) 'Kojima's theory of Japanese foreign direct investment revisited', *Hitotsubashi Journal of Economics*, Vol. 32, No. 2, pp. 103–109.

——. (1990) 'Problems and developments in the core theory of international business', *Journal of International Business Studies*, Fourth Quarter, pp. 657–665.

——. (1988) 'The limits of explanation: Testing the internalization theory of the multinational enterprise', *Journal of International Business Studies*, Summer, pp. 181–193.

——. (1983) 'New theories of international business: Some unresolved issues', in *The growth of international business*, edited by Mark Casson. London: George Allen&Unwin.

Buckley, Peter J. and Casson, Mark. (1991) *The future of the multinational enterprise*. London: Macmillan Press (Second edition).

Business Korea. (1992) *Yearbook on the Korean economy and business*. Seoul: Business Korea Co. Ltd.

Byrne, Thomas. (2000) 'Korea's rebound: A rating agency's perspective', in *Korea's economy 2000,* edited by the Korea Economic Institute Editorial Board. Washington DC: Korea Economic Institute of America.

Byun Jae-woong. (1995) *A new theoretical foreign direct investment model: The Korean foreign direct investment in the European Union as a case study*. Doctoral thesis. University of Leicester.

Calvet, A. L. (1981) 'A synthesis of FDI theories and theories of multinational firms', *Journal of International Business Studies*, 12, pp. 43–59.

Casson, Mark. (1992) 'Internalization theory and beyond', in *New directions in international business*, edited by Peter J. Buckley. Aldershot: Edward Elgar.

——. (1987) *The firm and the market: Studies on multinational enterprise and the growth of the firm*. Cambridge, Mass.: MIT Press

——. (1986) 'General theories of the multinational enterprise: Their relevance to business history', in *Multinationals: Theory and history*, edited by Peter Hertner and Geoffrey Jones. Aldershot: Gower.

——. (Ed). (1983) *The growth of international business*. London: George Allen&Unwin.

Caves, Richard E. (1982) *Multinational enterprise and economic analysis*. Cambridge: Cambridge University Press.

Cha Dong-se *et al.* (Ed.) (1997) *The Korean economy 1945–1995: Performance and vision for the 21st century*. Seoul: Korea Development Institute.

Chai Denise. (1993a) 'Hopes run high', *Business Korea*, July, pp. 37–45.

——. (1993b) 'To market, to market', *Business Korea*, May, pp. 28–29.

Chang Young-Jung. (1989) 'Economic relations between the European Community and East Asian economies', in *New dynamics in East-West relations*, edited by Kim D. C. *et al.* Seoul: Institute of East and West Studies, Yonsei University.

Cheong Inkyo. (1997) 'Europe-Korea investment relations', in *Europe-East Asia economic relations: Current status and prospects*, edited by Kim Bak-soo. Seoul: Korea Institute for International Economic Policy.

Chi Nam-ung. (1994) *Han'guk kajŏn sanŏb-ŭi kukchehwa kwajŏng-gwa chŏllyak* (The internationalization process and strategy of the Korean consumer electronics industry). Doctoral thesis. Seoul: Hankuk University of Foreign Studies Graduate School, Department of International Trade and Business.

Cho Dong-sung. (1998) 'Toward successful corporate restructuring', *Korea Focus*, Nov-Dec, Vol. 6, No. 6, pp. 17–24.

Cho Ki-ch'ang. (1996) *WTO shidae-ŭi haewoe chikchŏp t'uja hwaktae chŏllyak mit chŏngch'aek kwaje* (Policy issues and strategy for the expansion of foreign direct investment in the WTO era). Master's Degree dissertation. Seoul: Sŏgang University Graduate School of Economics, Department of International Economics.

Cho Lee-jay and Kim Yoon Hyung. (Ed.) (1998) *Korea's choices in emerging global competition and cooperation*. Seoul: Korea Development Institute.

Cho Lee-jay and Kim Yoon Hyung. (Ed.) (1995) *Economic systems in South and North Korea: The agenda for economic integration*. Seoul: Korea Development Institute.

Cho Soon. (1994) *The dynamics of Korean economic development*. Washington DC: Institute for International Economics.

Cho Tong-sŏng. (1989) *Kukche kyŏngnyŏnghak* (International business management). Seoul: Kyŏngmunsa.

Cho Wŏn-dong. (1993) 'Trade issues between Korea and the EC: Observation based on case study', in *The Asian Dimension of EC Integration: Problems and Prospects*, edited by Kim Bak-soo. Seoul: KIEP&Friedrich-Ebert-Stiftung Seminar, 10–11 November 1993.

Cho Yoon-je. (2000) 'Remaining challenges for financial restructuring in Korea', *Business Korea*, February, pp. 32–33.

——. (1999) 'Asian financial crisis and the future of Korean economy', *Korea Focus*, Jan-Feb, Vol. 7, No. 1, pp. 1–14.

Choi B. W. (1997) 'Economy likely to recover from September or October', *Economic Report*, September, pp. 30–31.

Choi Hochin. (1971) *The economic history of Korea*. Seoul: Sekyungsa.

Choi Jang-jip. (1998) 'Korea's political economy: Search for a solution', *Korea Focus*, Mar-Apr, Vol. 6, No. 2, pp. 1–20.

Choi Jinni. (1997a) 'Seoul, London see steady expansion of cooperation in broad areas', *Economic Report*, September, p. 72.

——. (1997b) 'Korean conglomerates vie with advanced country MNCs for global market shares', *Economic Report*, January, pp. 28–29.

Choi Yong-rok. (1996) *Haewoe t'ujaron: Iron-gwa shilmu* (Foreign investment theory: Theory and application). Seoul: Pagyŏngsa.

——. (1991) *Foreign direct investment by Korean manufacturing firms*. Doctoral thesis, University of Cincinnati.

Chŏng Hŏn-bae. (1995) 'EU shijang t'uja kiŏb-ŭi kyŏngjaeng uwi-wa hyŏnji chŏgŭng (The competitive advantage of firms investing in the EU and their

adaptation to the local area), in *Han'guk kiŏb-ŭi kukche kyŏngnyŏng chŏllyak* (The international management strategy of Korean firms), edited by Chŏng Ku-hyŏn and Lee Kwang-ch'ŏl. Seoul: Pŏmmunsa.
Chŏng Sŏng-ch'ŏl and Pak Tong-bae. (1992) *Han-EC-gan kwahak kisul hyŏmnyŏg-ŭi kuch'e pangan yŏn'gu* (A study of concrete plans for scientific and technological cooperation between Korea and the EC). Seoul: Han'guk Kwahak Kisul Yŏn'guwŏn.
Chŏng Yong-gil. (1983) 'Han-sŏgu kwan'gye ilsegi' (A century of Korean-Western European relations), *Kukche munje*, Vol. 14, No. 4, pp. 88–93.
Chung Jin-young. (1990) 'South Korean strategies for dynamic transformation: 1961–88' in *Dynamic transformation: Korea, NICs and beyond*, edited by Lim Gill-chin and Chang Wook. Illinois: Consortium on Development Studies.
Chung Oknim. (1998) 'Regional perspectives and roles on the Korean peninsula', *Korea and World Affairs*, Summer, Vol. 22, No. 2, pp. 222–241.
Chung Soon-won. (1996) 'Direction of economic policy for 1996', *Korea Focus,* January-February, Vol. 4, No. 1, pp. 62–67.
Chung Un-chan. (1997) 'A diagnosis of the Korean economy', *Korea Focus*, March-April, Vol. 5, No. 2, pp. 61–73.
Ch'a Wŏn-dae. (1997) *Han'guk chejo kiŏb-ŭi haewoe t'uja munjejŏm-gwa kaesŏn pangan: S chŏnja sarye-rŭl chungshim-ŭro* (A study of the problems of overseas investment by Korean manufacturing firms and a plan for its reform: The case study of S. Electronics). Doctoral thesis. Seoul: Sogang University, Graduate School of Business Administration.
Ch'ae Chŏm-gil. (1991) *Haewoe chikchŏp t'uja-ŭi hwalsŏnghwa pangan-e kwanhan yŏn'gu* (A study of plans to activate foreign direct investment). Doctoral thesis. Masan: Kyŏngnam University Graduate School, Department of Economics.
Ch'oi Hye-jin. (1997) *Ch'oigŭn haewoe chikchŏp t'uja ch'ui-e kwanhan yŏn'gu* (A study of recent developments in foreign direct investment). Master's Degree dissertation. Seoul: Hankuk University of Foreign Studies, Graduate School of Global Business.
Ch'oi Paeng-nyŏl. (1991) *Haewoe chikchŏp t'uja-ŭi chuyo kyŏltchŏng yoin-e kwanhan iron-gwa shiljŭng yŏn'gu: Kŏshijŏk kyŏngje pyŏnsu-rŭl chungshim-ŭro* (A theoretical and emprical study of the major determinants of foreign direct investment: A macroeconomic approach) Doctoral thesis. Chŏnju: Chŏnbuk National University Graduate School, Department of International Trade.
Claessens, Stijn. (2000) 'Korea's financial reform: Progress and challenges', in *Korea's economy 2000,* edited by the Korea Economic Institute Editorial Board, Washington DC: Korea Economic Institute of America.
Clifford, Mark L. (1994) *Troubled tiger: Businessmen, bureaucrats and generals in South Korea*. New York: M. E. Sharpe Inc.
Clough, Ralph N. (1990) 'North-South relations in Korea and implications for US policy', in *Korea-US relations in a changing world*, edited by Robert Sutter and Han Sungjoo. Berkeley: University of California, Institute of East Asian Studies.
Coase R. H. (1937) 'The nature of the firm', *Economica*, 4, pp. 386–405.
Crane, Geoff. (1993) 'So where's the recovery?', *Economic Report,* August, pp. 52–55.
Dent, Christopher M. (1999) *The European Union and East Asia: An economic relationship*. London: Routledge.
——. (1997) 'The ASEM: Managing the new framework of the EU's economic relations with East Asia', *Pacific Affairs*, Winter, Vol. 70, No. 4, pp. 495–516.
Dent, Christopher M. and Randerson, Claire. (1996) 'Korean foreign direct investment in Europe: the determining forces', *The Pacific Review*, Vol. 9, No. 4, pp. 531–552.

Department of Trade and Industry. (1995) *Britain and Korea to forge new partnership*. Press release, 9 March 1995.

Drennan, William M. (1998) 'The US role in Korean reunification', *Korea and World Affairs*, Summer, Vol. 22, No. 2, pp. 159–176.

Drysdale, Peter and Vines, David. (Ed.) (1998) *Europe, East Asia and APEC: A shared global agenda?* Cambridge: Cambridge University Press.

Dunning, John H. (1993a) *The globalization of business*. London: Routledge.

——. (1993b) *Multinational enterprises and the global economy*. Reading, Mass: Addison-Wesley.

——. (1988a) *Explaining international production*. London: Unwin Hyman Ltd.

——. (1988b) 'The eclectic paradigm of international production: A restatement and some possible extensions', *Journal of International Business Studies*, Spring, pp. 1–31.

——. (1987) 'The investment development cycle and third world multinationals' in *Multinationals of the south*, edited by Khushi M. Khan. London: Francis Pinter.

——. (1986) 'The investment cycle revisited', *Weltwirtschaftliches Archiv*, 122, pp. 667–677.

——. (1983) 'Changes in the level and structure of international production: the last one hundred years', in *The growth of international business*, edited by Mark Casson, London: George Allen and Unwin.

——. (1981a) 'Explaining outward direct investment of developing countries: In support of the eclectic theory of international production', in *Multinationals from developing countries*, edited by K. Kumar and Maxwell G. McLeod. Massachusetts: Lexington Books.

——. (1981b) *International production and the multinational enterprise*. London: George Allen&Unwin.

——. (1972) *International investment: Selected readings*. Harmondsworth: Penguin Modern Economics Readings.

Dunning, John H. and Rugman, A. M. (1985) 'The influence of Hymer's dissertation on the theory of foreign direct investment', *American Economic Review*, 75, May, 228–232.

Eberstadt, Nicholas. (1998) 'US-North Korea economic relations: Indications from North Korea's past trade performance', in *The US and the two Koreas: A new triangle*, edited by Park Tong Whan. Boulder, Colorado: Lynne Rienner Publishers Inc.

Eckert, Carter J. *et al.* (1990) *Korea old and new: A history*. Seoul: Ilchogak.

Electronics Industries Association of Korea. (1998) '1998 nyŏndo segye chŏnja shijang-ŭi hyŏnhwang-gwa chŏnmang' (The current status of and prospects for the world electronics industry in 1998), *Chŏnja chinhŭng*, February, pp. 5–48.

——. (1997a) *Chŏnja sanŏp tonghyang mit '98 nyŏn chŏnmang* (Trends in and prospects for the electronics industry in 1998). Seoul: EIAK.

——. (1997b) *Segye chŏnja sanŏb-esŏ-ŭi han'guk chŏnja sanŏp* (Korea's place in the global electronics industry). Seoul: EIAK.

——. (1997c) *Introduction to EIAK: A bridge connecting all of the electronics industries*. Seoul: EIAK.

Euh Yoon-dae and Min Sang H. (1989) 'Foreign direct investment from developing countries: The case of Korean firms', in *Management behind industrialization: Readings in Korean business*, edited by Kim Dong-ki and Kim Linsu. Seoul: Korea University Press.

Federation of Korean Industries. (1987) *Korea's economic policies (1945–1985)*. Seoul: Federation of Korean Industries.

Froot, Kenneth. (1993) *Foreign direct investment*. Chicago: University of Chicago Press (National Bureau of Economic Research Project Report).

Ghymn, Kyung-il. (1980) 'Multinational enterprises from the third world', *Journal of International Business Studies*, Fall, pp. 118–122.
Gittelman, Michelle and Dunning, John H. (1992) 'Japanese multinationals in Europe and the United States: Some comparisons and contrasts', in *Multinationals in the new Europe and global trade*, edited by Michael W. Klein and Paul J. J. Welfens. Berlin: Springer-Verlag.
Gomes-Casseres, Benjamin and Yoffie, David B. (1992) *The international political economy of direct foreign investment*. Aldershot: Edward Elgar Publishing Limited.
Graham, Edward M. and Krugman, Paul R. (1993) 'The surge in foreign direct investment in the 1980s', in *Foreign direct investment*, edited by Kenneth A. Froot. London: University of Chicago Press (National Bureau of Economic Research Project Report).
Gray, Peter H. (1996) 'The eclectic paradigm: The next generation', *Transnational Corporations*, Vol. 5, No. 2, pp. 51–65
Ha Hyŏn-bong. (1988) *Haewoe chikchŏp t'uja iron-ŭi ilgoch'al* (A survey of theories of foreign direct investment). Master's Degree dissertation. Seoul: Sŏnggyun'gwan University, Graduate School of Foreign Trade.
Ha Pyŏng-gi and Hong Sŏg-il. (1998) *Han'guk kiŏb-ŭi haewoe chikchŏp t'uja kyŏngnyŏng shilt'ae punsŏk* (An analysis of the actual management conditions of Korean investors overseas). Seoul: Sanŏp Yŏn'guwŏn.
Ha Sung-keun. (1998) 'Direction of financial sector restructuring', *Korea Focus*, Jul-Aug, Vol. 6, No. 4, pp. 35–43.
Ha Yong-ch'ŏl. (1994) *Dunning-ŭi cholch'ungiron-ŭro pon han'gug-ŭi haewoe chikchŏp t'uja: Tongnamasiagug-ŭl chungshim-ŭro* (Korea's foreign direct investment in Southeast Asia: An application of Dunning's Eclectic Theory of International Production). Master's Degree dissertation. Seoul: Sungshil University Graduate School, Department of Economics.
Hackney, Don. (1997) 'Here come the Europeans', *Korea Trade&Investment*, March-April, pp. 20–24.
Haggard, Stephan *et al.* (1994) *Macroeconomic policy and adjustment in Korea 1970–1990*. Harvard: Harvard Institute for International Development.
Halsey, Daniel. (1996) 'Going global: From rhetoric to reality', *Business Korea*, October, pp. 12–13.
Han Chong-su. (1998) *Yurŏp yŏnhap(EU)-kwa han'guk* (The EU and Korea). Seoul: Tongsŏngsa.
Han Sung-joo. (1996) 'ASEM is an opportunity for Korea', *Korea Focus*, Mar-Apr, Vol. 4, No. 2, pp. 145–146.
——. (1995). *Korea in a changing world: Democracy, diplomacy and future developments*. Seoul: Oreum Publishing House.
Han To-suk. (1996) *Uri nara-ŭi haewoe chikchŏp t'uja-wa kwanse chedo-ŭi kaesŏn panghyang* (Korean foreign direct investment and improvements to the taxation system). Seoul: Han'guk Chose Yŏn'guwŏn.
Han Woo-keun. (1970) *The history of Korea*. Seoul: Eul-Yoo Publishing Company.
Han Yŏng-ho. (1992) *TaeEC chikchŏp t'uja hwan'gyŏng pyŏnhwa-e ttarŭn taeŭng pangan yŏn'gu* (A study of the response to changes in the EC investment environment). Master's Degree dissertation. Pusan: Dong-A University, Department of Economic Graduate School.
Hang Chen. (1997) 'A slowdown, yes ... but not a crisis', *Business Korea*, September, p. 10.
Harding, Harry and Wang Jisi. (1998) 'The role of the United States', in *Europe and the Asia Pacific*, edited by Hanns Maull *et al.* London: Routledge.

Hayes, Peter and Noerper, Stephen. (1997) 'The future of the US-ROK alliance', in *Peace and security in Northeast Asia*, edited by Kihl Young Whan and Peter Hayes. New York: M. E. Sharpe.
Heenan, D. A. and Keegan, W. J. (1977) 'The rise of third world multinationals', *Harvard Business Review*, Jan-Feb, pp. 101–109.
Hennart, Jean-François. (1986) 'What is internalization?', *Weltwirtschaftliches Archiv*, 122, pp. 791–804.
Henthorn, William E. (1971) *A history of Korea*. New York: The Free Press.
Hirsch, S. (1976) 'An international trade and investment theory of the firm', *Oxford Economic Papers*, 28, pp. 258–270.
Hong Kyu-dok. (1999) 'North Korea-US relations proceeding too quickly', *Korea Focus*, Nov-Dec, Vol. 7, No. 6, pp. 116–118.
Hong, Mark. (1998) 'European-East Asian cooperation in international institutions', in *Europe and the Asia* Pacific, edited by Hanns Maull *et al.* London: Routledge.
Hong Sŏn-ch'ŏl. (1994) *Haewoe t'uja kiŏb-ŭi kyŏngjaeng uwi, t'ongje yuhyŏng mit kyŏngnyŏng sŏnggwa* (The competitive advantage of firms investing overseas, types of controls and management results). Master's Degree dissertation. Seoul: Sŏgang University Gradute School, Department of Trade.
Hong Sun-ho. (1995) 'Han'guk-kwa yurŏb-ŭi kwan'gye: Yŏng, tok, pur-ŭl chungshim-ŭro' (Korea-European relations: Focusing on England, Germany and France), *Woegyo*, September, pp. 80–91.
——. (1986) 'Chŏnhwan'gi-e sŏn han'guk-kwa yurŏp-kwa-ŭi kwan'gye' (ROK-European relations at a turning point), *T'ongil nonjip*, 1986, No. 6, Vol. 1, pp. 245–272.
Hong Sŭng-p'yo. (1990) *Uri nara chejŏb-ŭi haewoe t'uja shilt'ae-wa chŏllyag-e kwanhan yŏn'gu* (A study of the current status of and strategy regarding FDI by the Korean manufacturing industry). Master's Degree dissertation. Seoul: Yonsei University, Graduate Management School.
Hood, N. and Young, S. (1979) *The economics of multinational enterprise*. London and New York: Longman.
Horaguchi, Haruo and Toyne, Brian. (1990) 'Setting the record straight: Hymer, internalization theory and transaction cost economics', *Journal of International Business Studies*, Third Quarter, pp. 487–494.
Hur Nam-il. (1998a) 'Bouncing back', *Business Korea,* April, pp. 80–82.
——. (1998b) 'Korean companies on sale', *Business Korea*, February, pp. 22–23.
Hwang Myŏng-ju. (1993) *Han'guk kiŏb-ŭi haewoe chikchŏp t'uja-e kwanhan yŏn'gu* (A study of foreign direct investment by Korean firms). Master's Degree dissertation. Seoul: Kŏn'guk University Graduate School of Management, Department of Trade.
Hymer, Stephen Herbert. (1976) *The international operations of national firms: A study of direct foreign investment*. Cambridge, Mass. and London: MIT Press.
——. (1968) 'La grande corporation multinationale: Analyse de certain raisons qui poussent à l'intégration internationale des affaires'. *Révue Économique*, 14(6), pp. 949–973.
Hyun In-Taek. (1999) 'Seoul-Washington relations in the 21st century', *Korea Focus*, Nov-Dec, Vol. 7, No. 6, pp. 105–107.
Independent Task Force on Managing Change on the Korean Peninsula. (Ed.) (1998) *Managing change on the Korean peninsula: Report of an independent Task Force.* New York: Council on Foreign Relations Press.
Jeon Min-hee. (1997a) 'Exports begin to rebound', *Economic Report,* October, pp. 14–18.

——. (1997b) 'Government forges ahead with industrial restructuring', *Economic Report*, June, pp. 18–21.

Jeon Yoong-deok. (1992) 'The determinants of Korean foreign direct investment in manufacturing industries', *Weltwirtschaftliches Archiv*, Vol. 128, No. 3, pp. 527–541.

——. (1989) *The determinants of Korean foreign direct investment in the United States*. Doctoral thesis, Ohio State University.

Jeong Kap-young. (1999) 'Economic upswing amid "precarious stability"', *Korea Focus*, Jan-Feb, Vol. 7, No. 1, pp. 28–38.

Jeong Yeo-cheon. (1998) 'Recent developments and issues of Korea's economic relations with central and eastern European countries', in *Koreans and Korean business interests in Central Europe and CIS countries*, edited by Ahn Chung-si and Kwon Tai-hwan. Seoul: Seoul National University Press.

Jo Sung-hwan. (1981) 'Overseas direct investment by South Korean firms: Direction and pattern', in *Multinationals from developing countries*, edited by K. Kumar and Maxwell McLeod. Massachusetts: Lexington Books.

Johnson, H G. (1970) 'The efficiency and welfare implications of the international corporation', in *The international corporation*, edited by C. P. Kindleberger. Cambridge Mass: MIT Press. Reprinted in Dunning, John H. (Ed.) (1972) *International investment*. Harmondsworth, Penguin Books.

Jun W. T. And Prendergast, R. (1991) 'Upstream foreign direct investment by Korean manufacturers', in *Development perspectives for the 1990s*, edited by R. Prendergast and H. W. Singer. Basingstoke: Macmillan.

Jun Yongwook. (1989) *Kukchehwa shidae-ŭi haewoe t'uja chŏllyak* (Overseas investment strategy in the era of internationalization). Seoul: Sanŏp Yŏn'guwŏn.

——. (1988) *The structural analysis of the global consumer electronics industry and the oligopolistic behaviour of Korean firms in their internationalization*. Seoul: Korea Institute for Economics and Technology.

——. (1987) 'The reverse direct investment: The case of the Korean consumer electronics industry', *International Economic Journal*, Vol. 1, No. 3, pp. 99–102.

Jun Yongwook and Simon, Denis Fred. (1992) 'The pattern of Korea's foreign direct investment', in *Economic reform and internationalisation: China and the Pacific region*, edited by Ross Garnaut. St Leonards, New South Wales: Allen&Unwin.

Jun Yongwook and Yun Tong-jin (1991) 'Han'gug-ŭi taesŏnjin'guk chikchŏpt'uja-e taehan sogo: Kujojŏk shijangbangŏron' (An explanation of Korea's foreign direct investment in developed countries: Structural market defence approach), *Kukchekyŏngnyŏng yŏn'gu*, Vol. II, pp. 1–32.

Jung Ku-hyun. (1999) 'Asian economic crisis and corporate restructuring', *Korea Focus*, May-June, Vol. 7, No. 3, pp. 77–90.

——. (1991) 'Economic changes in Eastern Europe: An East Asian perspective', in *Europe in Transition and the Korean Peninsula*, edited by Kim D. C. *et al.* Seoul: Institute of East and West Studies, Yonsei University.

——. (1989) 'Direct investment and technology transfer between Korea and the European Community', in *New dynamics in East-West relations*, edited by Kim D. C. *et al.* Seoul: Institute of East and West Studies, Yonsei University.

Jwa Sung-Hee and Huh Chan Guk. (1998) 'Korea's 1997 currency crisis: Causes and implications', *Korea Journal*, Summer, Vol. 38, No. 2, pp. 5–33.

Kang Myung hun (1996). *The Korean business conglomerate: Chaebol then and now*. Berkeley: University of California, Institute of East Asian Studies.

Kang Sang-yun. (1994) 'Businesses may see higher interest rates and prices', *Business Korea*, December, pp. 24–25.

Kang Suk-hŭi. (1986) *Mi, yurŏp, il, han'gug-ŭi haewoe chikchŏp t'uja-e taehan pigyo yŏn'gu* (A comparative study of American, European, Japanese and Korean foreign direct investment). Master's Degree dissertation. Seoul: Sungmyŏng University Graduate School, Department of International Trade.

Kang T'ae-gu. (1997) *Tagukchŏk kiŏp kyŏngnyŏngnon* (The theory of multinational enterprise management). Seoul: Muyŏk Kyŏngnyŏnsa.

Khan, Khushi M. (1987) 'Multinationals from the south: Emergence,patterns and issue', in *Multinationals of the south*, edited by Khushi M. Khan. London: Francis Pinter.

——. (Ed.) (1987) *Multinationals of the south*. London: Francis Pinter.

Kiernan, Terrence Gerard. (1995) 'Destination: An open economy', *Economic Report*, March, pp. 34–35.

——. (1994a) 'Korean déjà vu?', *Economic Report,* September, pp. 18–20.

——. (1994b) 'Back on track', *Economic Report,* July, pp. 20–21.

——. (1994c) 'Outward bound', *Economic Report,* June, pp. 26–27.

——. (1993) 'Oh no, not again!', *Korea Business World,* November, p. 16.

Kihl Young Whan. (1997) 'Confrontation or compromise? Lessons from the 1994 crisis', in *Peace and security in Northeast Asia*, edited by Kihl Young Whan and Peter Hayes. New York: M. E. Sharpe.

——. (Ed.) (1994) *Korea and the world*. Boulder, Colorado:Westview Press.

Kihl Young Whan and Hayes, Peter. (Ed.) (1997) *Peace and security in Northeast Asia*. New York: M. E. Sharpe.

Kim Bak-soo. (Ed.) (1994) *Europe-East Asia economic relations: Current status and prospects*. Seoul: Korea Institute for International Economic Policy.

——. (1993) 'Changes in trade structure between the EC and Korea', in *The Asian Dimension of EC Integration: Problems and Prospects*, edited by Kim Bak-soo. Seoul: KIEP&Friedrich-Ebert-Stiftung Seminar, 10–11 November 1993.

——. (1992) *Pŏmyurŏp kyŏngjegwŏn hyŏngsŏng-ŭi chŏnmang-gwa yŏnghyang* (The prospects for and influence of the formation of a pan-European bloc). Seoul: Taewoe Kyŏngje Chŏngch'aek Yŏn'guwŏn.

Kim Chin-yŏng. (1990) *Uri nara kajŏn chep'um kiŏb-esŏ-ŭi ch'okchin chŏllyag-e kwanhan yŏn'gu: D-sa-ŭi sarye-rŭl chungshim-ŭro* (A study of the promotional strategy of electronic home appliance companies: The case study of D. Company). Master's Degree dissertation. Seoul: Sogang University, Graduate School of Business Administration.

Kim Chŏng-guk. (1988) *Uri nara haewoe t'uja-ŭi hwaktae pangan-e kwanhan yŏn'gu: Haewoe chikchŏp t'uja-rŭl chungshim-ŭro* (The expansion of Korean foreign investment, centering on foreign direct investment). Master's Degree dissertation. Seoul: Sŏnggyun'gwan University Graduate School, Department of International Business.

Kim Chŏng-nyul. (1994) *EC shijang t'onghab-e taeŭnghan uri nara kiŏb-ŭi tae- EC chikchŏp t'uja hwaktae pangan-e kwanhan yŏn'gu: Ilbon kiŏp chinch'ul chŏllyak-kwa-ŭi pigyo-rŭl chungshim-ŭro* (A study of the effective strategies of Korean companies' foreign direct investment to cope with the integration of the EC market: A comparison with the entry strategies of Japanese firms). Master's Degree dissertation. Seoul: Sogang University, International Management School.

Kim Chong-tae. (1999a) 'Korea swings door wide open', *Business Korea*, June, pp. 26–27.

——. (1999b) 'Policy shifts toward growth', *Business Korea*, May, pp. 24–26.

——. (1999c) 'Heavy debt remains long-term burden', *Business Korea*, April, pp. 26–27.

——. (1999d) 'Recovery will be slow and painful', *Business Korea*, March, pp. 14–17.

——. (1998a) 'For sale: Korea Inc.', *Business Korea*, December, pp. 14–16.
——. (1998b) 'Free-market principles sacrificed for big deals', *Business Korea*, October, pp. 14–15.
——. (1998c) 'Quick corporate reform vital to economy', *Business Korea*, September, p. 14.
——. (1998d) 'Hoping the worst is over', *Business Korea*, September, pp. 56–57.
——. (1998e) 'Survival instinct drives sudden urge to merge', *Business Korea*, August, pp. 14–15.
——. (1998f) 'Red ink puts several banks on blacklist', *Business Korea*, August, pp. 17–18.
——. (1998g) 'Welcome, foreign buyers', *Business Korea*, July, pp. 20–24.
——. (1998h) 'Hot summer for bankers', *Business Korea*, July, pp. 50–51.
——. (1998i) 'Changing of the guard?', *Business Korea*, June, pp. 12–13.
——. (1998j) 'Korea's financial industry is ready for reform', *Business Korea*, April, pp. 24–27.
——. (1998k) 'Corporate Korea on sale', *Business Korea*, April, p. 49.
——. (1998l) 'Economic democracy', *Business Korea*, March, pp. 30–31.
——. (1998m) 'Are the good times over?', *Business Korea*, February, pp. 16–19.
——. (1998n) 'End of the crunch?', *Business Korea*, February, pp. 50–51.
——. (1998o) 'Surgery for the chaebols', *Business Korea*, January, pp. 34–35.
——. (1997a) 'The party's over', *Business Korea,* December, pp. 18–21.
——. (1997b) 'Excessive expansion and too much debt', *Business Korea,* November, pp. 16–17.
——. (1997c) 'Critical condition?', *Business Korea*, September, pp. 20–21.
——. (1997d) 'Falling farther and farther behind', *Business Korea,* July, pp. 14–16.
——. (1997e) 'The domino effect', *Business Korea*, June, pp. 16–17.
——. (1997f) 'Signs of recovery', *Business Korea*, June, pp. 14–15.
——. (1997g) 'Does anyone have a magic wand?', *Business Korea*, April, pp. 14–17.
——. (1997h) 'Korean products hit hard by falling Yen', *Business Korea*, March, pp. 16–17.
——. (1997i) 'Stagflation ahead', *Business Korea*, January, p. 18.
——. (1997j) 'There's a light at the end of the tunnel', *Business Korea,* January, p. 20.
——. (1996a) 'It's time to play in the major leagues', *Business Korea*, November, pp. 714–15.
——. (1996b) 'Electronics industry: Tough challenges ahead', *Business Korea*, November, pp. 18–19.
——. (1996c) 'Complex cure for coming doldrums', *Economic Report,* August, pp. 42–43.
——. (1996d) 'Storm on the horizon', *Economic Report*, July, pp. 22–24.
——. (1995a) 'Economy in high gear', *Economic Report,* April, pp. 22–24.
——. (1995b) 'A big pill to swallow', *Economic Report*, March, pp. 29–32.
——. (1994) 'Boom days', *Economic Report,* November, pp. 32–33.
——. (1993) 'Brighter prospects'. *Economic Report,* December, pp. 31–33.
Kim Ch'ŏl. (1998) *Haewoe t'ujaron* (The theory of foreign direct investment). Seoul: Tŭnam.
——. (1996) *Tagukchŏk kiŏmnon* (The theory of the multinational enterprise). Seoul: Pŏmmunsa.
Kim Chung-soo. (1997) 'Fading myth of chaebol's infallibility', *Korea Focus,* Sep-Oct, pp. 50–62.
Kim Citi. (1999a) 'Problems will not go away, Moody's warns', *Business Korea*, June, pp. 19–20.

——. (1999b) 'Joblessness likely to become chronic', *Business Korea*, April, pp. 16–17.

——. (1999c) 'Ready or not, foreign exchange will be deregulated', *Business Korea*, March, p. 11.

——. (1998a) 'Conglomerates defy downsizing initiative', *Business Korea*, December, pp. 20–21.

——. (1998b) 'Massive pump priming is on the way', *Business Korea*, November, pp. 12–13.

——. (1998c) 'More pain, not gain, on the way', *Business Korea*, November, pp. 27–28.

Kim D. C. *et al.* (Ed.) (1989) *New dynamics in East-West relations*. Seoul: Institute of East and West Studies, Yonsei University.

Kim E. Han. (1998) 'The Korean financial crisis and the future of its economy', *Korea Journal*, Summer, Vol. 38, No. 2, pp. 56–65.

Kim Hakjoon. (1993). *Korea's relations with her neighbours in a changing world.* Seoul*: Hollym.*

Kim Hong-ki. (1986) 'Early Korea-British relations', in *Korea and Britain tomorrow: Towards a broader relationship*, edited by Chung Chong-wha and Boyd McCleary. Seoul: Anglo-American Studies Institute, Korea University.

Kim Hordon. (1996) 'Offshore production settles down for local sales', *Business Korea,* April, pp. 47–48.

Kim Inchul. (1997) 'Trade deficit reduction high policy priority', *Economic Report*, January, pp. 18–19.

Kim In-june. (1998) 'One year under the IMF', *Korea Focus*, Nov-Dec, Vol. 6, No. 6, pp. 1–16.

Kim, James. (1998) 'Government urges massive M&A', *Business Korea*, September, pp. 44–45.

Kim Ji-hong. (1990) 'The structure of Korea's trade with the EC countries', in *The single European market and its implications for Korea as an NIE*, edited by Young Soogil and Kang Moonsoo. Seoul: Korea Development Institute & Friedrich-Ebert-Stiftung Conference, 18–19 June, 1990.

Kim June-dong. (2000) 'Foreign direct investment: Trends and policies', in *Korea's economy 2000,* edited by the Korea Economic Institute Editorial Board. Washington DC: Korea Economic Institute of America.

Kim Ki-jung. (1987) *Haewoe chikchŏp t'uja-ŭi iron-gwa hyŏngt'ae-e kwanhan yŏn'gu* (A study of the theory and types of foreign direct investment). Master's Degree dissertation. Seoul: Hanyang University, Graduate School of Business Administration.

Kim Kwang Suk and Kim Joon-Kyung.(1997) 'Korean economic development: An overview', in *The Korean economy 1945–1995: Performance and vision for the 21st century*, edited by Cha Dong-se *et al.* Seoul: Korea Development Institute.

Kim Kwang-suk and Park Joon-kyung (1985). *Sources of economic growth in Korea: 1963–1982*. Seoul: Korea Development Institute.

Kim Kyŏng-ho. (1989) *Uri nara-ŭi haewoe t'uja chedo mit chŏngch'aeg-e kwanhan yŏn'gu* (A study of Korea's overseas investment system and policy). Master's Degree dissertation. Seoul: Kyunghee University Graduate School, Department of Economics.

Kim Kyu-ch'ang. (1984) *Haewoe chikchŏp t'ujaron: T'uja taesang chiyŏg-ŭi hwan'gyŏng yoin punsŏk p'yŏngga* (The theory of foreign direct investment: An analystical evaluation of host region environmental factors). Seoul: Pŏmmunsa.

Kim Kyu-dong. (1994) *Yurŏp kongdongch'e-e taehan uri chŏngbu-ŭi chŏngch'aekchŏk taeŭng* (The Korean government's policy response regarding the European

Union). Master's Degree dissertation. Seoul: Hankuk University of Foreign Studies, Department of Public Administration Graduate School.

Kim Kyu-p'an. (1995) *Tonggu-ŭi kyŏngje yŏgŏn pyŏnhwa-wa hanŏtonggu kyŏnghyŏp chŭngjin pangan* (Changes in the economic circumstances of eastern Europe and proposals for the advancement of economic cooperation between Korea and eastern Europe). Seoul: Taewoe Kyŏngje Chŏngch'aek Yŏn'guwŏn.

Kim Kyung-ho. (1996) 'Korea hopes to be mediator between Asian, EU blocs', *Korea Newsreview,* 2 March, p. 10.

Kim Kywon. (2000) 'Chaebol reform: Rationale and contradictions', *Korea Focus*, Jan-Feb, Vol. 8, No. 1, pp. 74–91.

Kim Linsu and Seong So-mi. (1997) 'Science and technology: Public policy and private strategy', in *The Korean economy 1945–1995: Performance and vision for the 21st century*, edited by Cha Dong-se *et al.* Seoul: Korea Development Institute.

Kim S. C. (1996) 'Foreign investment ushers in new phase of international trade relations', *Business Korea,* May, pp. 26–29.

Kim, Samuel S. (1994) 'The two Koreas and world order', in *Korea and the world*, edited by Kihl Young Hwan. Boulder, Colorado: Westview Press.

Kim Sang Kyom. (1997) 'Europe-East Asia trade relations: An East Asian perspective', in *Europe-East Asia economic relations: Current status and prospects,* edited by Kim Bak-soo. Seoul: Korea Institute for International Economic Policy.

Kim Seon-Young. (1998) 'Removing the stumbling blocks', *Business Korea*, May, pp. 32–33.

Kim Seong-soo. (1993) *Determinants of the magnitude of foreign direct investment: An analysis of Korean manufacturing MNCs*. Doctoral thesis. University of North Texas.

Kim Seung-chuan. (1999) 'Seoul government determined to reduce chaebol control of financial industry', *Economic Report*, July, p. 39.

Kim Shi-jung *et al.* (1992) *Haewoe chikchŏp t'uja-ŭi hyŏnhwang-gwa chŏngch'aek munje* (The current status of and policy issues in foreign direct investment). Seoul: Taewoe Kyŏngje Chŏngch'aek Yŏn'guwŏn.

Kim Shin. (Ed.) (1992) *Tagukchŏk kiŏp yŏn'gu* (A study of the multinational enterprise). Seoul: Sŏkchŏn.

Kim Sŏk-hŭi *et al.* (1989) *Kungnae chŏnja sanŏb-ŭi issŏsŏ-ŭi woegugin chikchŏp t'uja-ŭi hyogwa* (The effects of foreign direct investment in the Korean electronics industry). Seoul: Sanŏp Yŏnguwŏn.

Kim Sŭng-jin. (1998) *Haewoe chikchŏp t'uja-ŭi yoin-gwa hyogwa punsŏk* (An analysis of the key factors in and effects of foreign direct investment). Seoul: Han'guk Kaebal Yŏn'guwŏn.

Kim Tong-su. (1980) *Tagukchŏk kiŏb-e kwanhan ironjŏk chŏpkŭn-gwa shilt'ae punsŏk* (A theoretical approach towards and empirical analysis of multinational corporations). Master's Degree dissertation. Seoul: Kŏn'guk University Graduate School.

Kim Yong-ho. (1998) 'Status and problems of Korean enterprises advancing into central and eastern Europe', in *Koreans and Korean business interests in Central Europe and CIS countries*, edited by Ahn Chung-si and Kwon Tai-hwan. Seoul: Seoul National University Press.

Kim Young-tai. (1999) 'Rising voices of concern over the economy', *Korea Focus*, Mar-Apr, Vol. 7, No. 2, pp. 65–70.

Kim Yu-hwan. (1984) *Han'gug-ŭi haewoe t'uja chiwŏn chedo-e kwanhan yŏn'gu* (A study of Korea's foreign direct investment support systems). Master's Degree dissertation. Seoul: Chungang University, International Management School.

Kim Yei-tae. (1995) 'Nation exports production bases', *Economic Report,* October, pp. 12–13.

Kindleberger, Charles P. (1984) *Multinational excursions*. Cambridge MA: MIT Press.

——. (1969) 'The theory of direct investment', in *American business abroad*, edited by Charles P. Kindleberger. New Haven, London: Yale University Press.

Klein, Michael W. and Welfens, Paul J. J. (Ed.) (1992) *Multinationals in the new Europe and global trade*. Berlin: Springer-Verlag.

Knickerbocker, F. T. (1973) *Oligopolistic reaction and the multinational enterprise*. Boston: Harvard University Press.

Kojima Kiyoshi. (1989) 'Theory of internalisation by multinational corporations' *Hitotsubashi Journal of Economics*, Vol. 30, pp. 65–85.

Kojima Kiyoshi. (1986) 'Japanese-style direct foreign investment', *Japanese Economic Studies*, Spring. Reprinted in *The international political economy of direct foreign investment*, edited by Benjamin Gomes-Casseres and David B. (1992). Aldershot: Edward Elgar Publishing Limited.

——. (1978) *Direct foreign investment: A Japanese model of multinational business operations*. London: Croom Helm.

——. (1977) 'Direct foreign investment between advanced industrialized countries', *Hitotsubashi Journal of Economics*, June, pp. 1–18.

——. (1973) 'A macroeconomic approach to foreign direct investment', *Hitotsubashi Journal of Economics*, June, pp. 1–21.

Kojima Kiyoshi and Ozawa Terutomo. (1984) 'Micro- and macro-economic models of direct foreign investment: Toward a synthesis', *Hitotsubashi Journal of Economics*, Vol. 25, No. 2, pp. 1–20.

Koo Bon Ho and Jo Dongho. (1995) 'Comparative analysis of the North and South Korean economies', in *Economic systems in South and North Korea: The agenda for economic integration*, edited by Cho Lee-jay and Kim Yoon Hyung. Seoul: Korea Development Institute.

Koo Bohn-young. (1985) 'Korea', in *Multinational enterprises, economic structure and international competitiveness*, edited by John H. Dunning. Chichester: John Wiley and Sons.

Koo Y. K. (1996) 'Korea's new economic crisis?', *Business Korea*, October, p. 64.

Koopmann, Georg and Scharrer, Hans-Eckart. (1990) 'Scenarios of a common external trade policy for the EC after 1992', in *The single European market and its implications for Korea as an NIE*, edited by Young Soogil and Kang Moonsoo. Seoul: Korea Development Institute & Friedrich-Ebert-Stiftung Conference, 18–19 June, 1990.

Korea Development Institute. (1986) *Korea Year 2000: Prospects and issues for long-term development*. Seoul: KDI Press.

Korea Economic Institute Editorial Board. (Ed.) (2000) *Korea's economy 2000.* Washington DC: Korea Economic Institute of America.

Korea Overseas Information Service. (1993) *Facts about Korea*. Seoul: Korea Overseas Information Service.

Kreisberg, Paul H. (1990) 'US-Korea political relations: Emerging policy issues', in *Korea-US relations in a changing world*, edited by Robert Sutter and Han Sungjoo. Berkeley: University of California, Institute of East Asian Studies.

Ku Cha-yong. (1992) *Han'guk kajŏn sanŏb-ŭi segyehwa-rŭl wihan haewoe chikchŏp t'uja chŏllyag-e kwanhan yŏn'gu* (A study of foreign direct investment by and the globalization strategy of the Korean consumer electronics industry). Master's Degree dissertation. Seoul: Chungang University, Graduate School of International Management.

Kum, Kevin. (1997a) 'Electronics industry overview: Recovery ahead', *Business Korea*, October, pp. 33–34.

——. (1997b) 'Consumer electronics: Domestic products in a slump', *Business Korea,* July, p. 29.

Kumar, Krishna and Kim K. Y. (1984) 'The Korean manufacturing multinationals', *Journal of International Business Studies*, 15, pp. 45–62.

Kumar, Krishna and McLeod, Maxwell G. (1981). *Multinationals from developing countries*. Lexington, Mass. and Toronto: Lexington Books.

Kwak Young-sup. (1999) 'Dried up funds: Korean firms slash direct investment abroad', *Newsreview*, 15 May, pp. 26–27.

Kwon Chang-hee. (1997) 'Hanbo Iron & Steel fails to honor bills, goes bankrupt', *Economic Report*, February, pp. 18–19.

Kwŏn Chŏng-nyŏl. (1984) *Tagukchŏk kiŏmnon: Chŏllyak-kwa chinch'ul* (The theory of the multinational enterprise: Strategy and advance). Seoul: Hangmunsa.

Kwon, O. Yul. (2000) 'The Korean financial crisis and international business', *Asian Studies Review*, Vol. 24, No. 1, March, pp. 25–50.

Lall, Sanjaya. (1983) *The new multinationals: The spread of third world enterprises*. Chichester: John Wiley & Sons.

Lall, Sanjaya and Streeten, P. (1977) *Foreign investment, transnationals and developing countries*. London: Macmillan Press.

Langhammer, Rolf. (1998) 'Europe's trade, investment and strategic policy interests in Asia and APEC', in *Europe, East Asia and APEC: A shared global agenda?*, edited by Peter Drysdale and David Vines. Cambridge: Cambridge University Press.

——. (1990) 'European Integration and its impact on future economic relations with Korea: Where do we go in the 1990s', in *The single European market and its implications for Korea as an NIE*, edited by Young Soogil and Kang Moonsoo. Seoul: Korea Development Institute & Friedrich-Ebert-Stiftung Conference, 18–19 June, 1990.

Lecraw, Donald J. (1981) 'Internationalization of firms from LDCs: Evidence from the ASEAN region', in *Multinationals from developing countries*, edited by Krishna Kumar and Maxwell G. McLeod. Lexington, Mass. and Toronto: Lexington Books.

——. (1977) 'Direct investment by firms from less developed countries', *Oxford Economic Papers*, 29, No. 3.

Lee Byung-jong. (1995) 'In pursuit of stability', *Business Korea*, January, pp. 23–25.

——. (1994) 'Electronics industry: New whiz kids on the bloc' *Business Korea*, March, pp. 24–26.

Lee Chae-jin. (1994) 'The United States and Korea: Dynamics of changing relations', in *Korea and the world*, edited by Kihl Young Hwan. Boulder, Colorado: Westview Press.

Lee Chae-min. (1997) 'Uri nara haewoe chikchŏp t'uja-ŭi t'ŭkching mit sŏnggwa punsŏk' (The characteristics and an analysis of the results of Korean foreign direct investment), *Suŭn chosa wŏlbo*, October, pp. 4–34.

Lee Chang-jae. (1993) 'East Asian economy: Forces shaping Korea's politico-economic development', in *Regional context for Korea's dynamism: Certainties vs uncertainties*, edited by Lee Chang-jae. Seoul: Korea Institute for International Economic Policy.

——. (Ed.) (1993) *Regional context for Korea's dynamism: Certainties vs uncertainties*. Seoul: Korea Institute for International Economic Policy.

Lee Chong-hwa. (1997) *Han-EU chuyo t'ongsang hyŏnan-gwa taeŭng pangan* (Current key issues on Korea-EU trade disputes and policy recommendations). Seoul: Taewoe Kyŏngje Chŏngch'aek Yŏn'guwŏn.

Lee Chong-uk. (1992) *Han'guk chŏnja sanŏb-ŭi kyŏngjaengnyŏk wŏnch'ŏn-e kwanhan yŏn'gu* (A study of the sources of competitiveness in the Korean electronics industry). Seoul: Han'guk Kyŏngje Yŏn'guwŏn.
Lee Ch'an-gu. (1984) *Haewoe chikchŏp t'uja-wa tagukchŏk kiŏmnon* (The theory of foreign direct investment and the multinational enterprise). Seoul: Ek'onomia.
Lee E. W. (1997) 'Economy to grow 6.4% in 1997', *Economic Report,* November, pp. 22–23.
Lee Eu-chul. (1996) 'Exodus of Korean conglomerates', *Korea Focus,* September-October, Vol. 4, No. 5, pp. 75–88.
Lee Han-ku. (1995) 'Korea's economy will slow down after the end of this year', *Business Korea*, September, pp. 8–9.
Lee Hong-gue. (1996) *Challenges and responses in the new trading environment: A Korean perspective*. Seoul: Korea Institute for International Economic Policy.
——. (1990) 'EC 1992 and Korea's direct investment in the EC', in *The single European market and its implications for Korea as an NIE*, edited by Young Soogil and Kang Moonsoo. Seoul: Korea Development Institute & Friedrich-Ebert-Stiftung Conference, 18–19 June, 1990.
Lee Hyung-koo. (1996) *The Korean economy: Perspectives for the twenty-first century*. Albany: State University of New York Press.
Lee Kap-soo. (1997) 'IMF assistance only way to overcome current foreign exchange crisis', *Newsreview,* 22 November, p. 16.
Lee Keun-yung. (2000) 'The future for the Korean economy and the transition to a new development paradigm', in *Korea's economy 2000,* edited by the Korea Economic Institute Editorial Board. Washington DC: Korea Economic Institute of America.
Lee Ki-dong. (2000) 'Corporate thinking embraces new economy', *Business Korea*, April, pp. 46–47.
Lee Kyong-hwan. (1998) 'Speeding up reform', *Business Korea*, May, pp. 30–31.
Lee Nak-ho. (1998) 'S. Korea, IMF ink US$57 billion bailout loan plan', *Economic Report,* January, pp. 33–34.
——. (1997) 'S. Korea seeks loans from IMF', *Economic Report,* December, pp. 42–43.
Lee Sahng-gyoun. (1999) 'ASEM and a new world order: Challenges and prospects', *Korea Focus*, Sep-Oct, Vol. 7, No. 5, pp. 20–33.
Lee Sang-Gon and Ruffini, Pierre-Bruno. (1999) *The global integration of Europe and East Asia.* Cheltenham: Edward Elgar.
Lee Sang-ho. (1988) *Uri nara kiŏb-ŭi haewoe chikchŏp t'uja-e kwanhan yŏn'gu* (A study of Korean foreign direct investment). Master's Degree dissertation. Taegu: Yŏngnam University Graduate School, Department of Trade.
Lee So-Yon. (1999) 'Recession lingers for much of economy', *Business Korea*, May, p. 22.
——. (1998) 'Korea is on a roll, relatively speaking', *Business Korea*, December, pp. 40–43.
Lee Sŏn-sang. (1989) *Han'guk kajŏn kiŏb-ŭi haewoe t'uja chŏllyag-e taehan yŏn'gu: Ssa-ŭi CTV haewoe saengsan-ŭl chungshim-ŭro* (A study of the foreign investment strategy of Korean home electronics companies: The overseas production of colour televisions by S. company). Master's Degree dissertation. Seoul: Yonsei University, Graduate School of Business Administration.
Lee Tai Hwan. (1999) 'Korea's foreign policy in the post-cold war era', *Korea Focus,* Nov-Dec, Vol. 7, No. 6, pp. 1–17.
Lee Ŭng-gwŏn. (1993) *Haewoe chikchŏp t'uja-ŭi mohyŏng sŏljŏng mit kyoran yoin-e kwanhan yŏn'gu: Uri nara chejoŏb-ŭi tong asia haewoe chikchŏp t'uja chungshim-ŭro* (A study of the model shaping and determinants for foreign direct

investment: Korean manufacturing foreign direct investment in East Asia). Doctoral thesis. Seoul: Sejong University Graduate School, Department of Trade.
Lee Yeon-ho. (1997) *The state, society and big business in South Korea*. London: Routledge.
Lee Yeon Ho and Choi Yearn Hong. (1995) 'The government-business (chaebol) relationship in South Korea: 1988–1994', *Asian Thought and Society*, Vol. 20, No. 58–59, pp. 93–113.
Lee Yŏng-su. (1990) *Han'guk kiŏb-ŭi tae-EC chikchŏp t'uja chŏllyag-e kwanhan yŏ-n'gu* (A study of Korean enterprises' direct investment strategy in the EC). Master's Degree dissertation. Seoul: Sŏnggyun'gwan University Graduate School of Trade, Department of Trade.
Lie, John. (1998) *Han unbound: The political economy of South Korea*. Stanford: Stanford University Press.
Lieberman, Ira W. and Mako, William P. (2000) 'Corporate restructuring in Korea', in *Korea's economy 2000*, edited by the Korea Economic Institute Editorial Board. Washington DC: Korea Economic Institute of America.
Lim, Gill-chin and Chang Wook. (Ed.) (1990) *Dynamic transformation: Korea, NICs and beyond*. Illinois: Consortium on Development Studies.
Lim, Herman Eng Hock. (1993) *A study into the growth, style and direct overseas investment by Korean conglomerates.* M.B.A. Dissertation. Sheffield University.
Liu, Scott X. (1997) *Foreign direct investment and the multinational enterprise: A re-examination using signaling theory*. Westport: Praeger Publishers.
Lone, Stewart and McCormack, Gavan. (1993) *Korea since 1850*. New York: St Martin's Press.
Macdonald, Donald Stone. (1990) *The Koreans: Contemporary politics and society*. Colorado: Westview Press Inc.
Magee, S. P. (1976) 'Technology and the appropriability theory of the multinational corporation', in *The new international economic order*, edited by Jagdish Bhajwati. Cambridge MA: MIT Press.
Maitland, Donald and Hu Yao-su. (Ed.) (1999)*Europe and emerging Asia*. London: Federal Trust.
Mason, E. S. *et al.* (1980) *The economic and social modernization of the Republic of Korea*. Cambridge, Mass: Harvard University Press.
Mason R. Hal. (1980) 'A comment on Professor Kojima's Japanese type versus American type of technology transfer', *Hitotsubashi Journal of Economics*, Vol. 20, pp. 42–52.
Massman, Jason. (1993) 'Bottoming out', *Korea Business World*, December, p. 20.
Maull, Hanns *et al.* (Ed.) (1998) *Europe and the Asia Pacific*. London: Routledge.
McCulloch, Rachel. (1993) 'New perspectives on foreign direct investment', in *Foreign direct investment*, edited by Kenneth A. Froot. London: University of Chicago Press (National Bureau of Economic Research Project Report).
McGonagle, Paul. (2000) 'Korea's new economy sailing ahead', in *Korea's economy 2000*, edited by the Korea Economic Institute Editorial Board. Washington DC: Korea Economic Institute of America.
McNamara, Dennis L. (1990) *The colonial origins of Korean enterprise: 1910–1945*. Cambridge: Cambridge University Press.
Menon, Jayant and Athukorala, Premachandra. (1995) 'Korean foreign direct investment: Trends, determinants and prospects', *World competition: Law and economics*, Vol. 18 (4), pp. 55–65.
Min Chung-Ki. (1993) *EC kyŏngjaeng chŏngch'aek pyŏnhwa-wa uri-ŭi taeŭng pangan* (Changes in EC competition policy and our response). Seoul: Taewoe Kyŏngje Chŏngch'aek Yŏn'guwŏn.

——. (1991a) 'A framework for a better relationship between Korea and the European Community', in *Europe in Transition and the Korean Peninsula*, edited by Kim D. C. *et al.* Seoul: Institute of East and West Studies, Yonsei University.

——. (1991b) 'Korean direct investment in the European Community: Strategies for Korean firms', in *The economic cooperation between EC and Korea: Problems and prospects,*edited by Min Chung-ki. Seoul: KIEP & Friedrich-Ebert- Stiftung Seminar, 14–15 October 1991.

——. (Ed.) (1991c) *The economic cooperation between EC and Korea: Problems and prospects.* Seoul: KIEP & Friedrich-Ebert-Stiftung Seminar, 14–15 October 1991.

——. (1991d) *EC-ŭi chikchŏp t'uja hwan'gyŏng pyŏnhwa-wa taeŭng pangan* (Changes in the EC investment environment and plans for a response). Seoul: Taewoe Kyŏngje Chŏngch'aek Yŏn'guwŏn.

——. (Ed.) (1991e) *EC t'onghap-kwa uri-ŭi taeŭng pangan* (EC integration and our response). Seoul: Taewoe Kyŏngje Chŏngch'aek Yŏn'guwŏn.

Montagnon, Peter and Burton, John. (1999) 'Pruning the chaebol', *Financial Times*, September 23, p. 15.

Nahm, Andrew C. (1988) *Korea: Tradition and transformation – A history of the Korean people*. Seoul: Hollym.

Nam In-soo. (1998) 'Chaebol urged to cut core businesses to 3–4', *Newsreview,* 21 February, p. 9.

Nam Sang-kun and Slater, Jim. (1997) 'Korean investment in Europe: Motives and choices', in *Business relationships with East Asia: The European experience*, edited by Jim Slater and Roger Strange. London: Routledge.

No T'aek-hwan. (1993) *Tagukchŏk kiŏb-ŭi haewoe chikchŏp t'uja iron* (The theory of foreign direct investment by multinational enterprises). Seoul: Hangmunsa.

Nuttall, Simon. (1998) 'Europe and Northeast Asia', in *Europe and the Asia Pacific*, edited by Hanns Maull *et al.* London: Routledge.

Oh Ingyu. (1997) *Making an octopus.* Doctoral thesis. University of Oregon.

Oh Jong-nam. (1997) 'Europe-East Asia economic relations: Policy challenges: Korean perspective', in *Europe-East Asia economic relations: Current status and prospects*, edited by Kim Bak-soo. Seoul: Korea Institute for International Economic Policy.

Ŏm Ch'ang-sŏp. (1991) 'EC t'onghap taebi t'ongsang sanŏp pumun taeŭng chŏllyak' (The response of the trade sector to EC integration), in *EC t'onghap- kwa uri-ŭi taeŭng pangan* (EC integration and our response), edited by Min Chung-ki. Seoul: Taewoe Kyŏngje Chŏngch'aek Yŏn'guwŏn.

Organization for Economic Cooperation and Development. (1992) *International direct investment: Policies and trends in the 1980s*. Paris: OECD.

Ozawa Terutomo. (1985) 'Japan', in *Multinational enterprises, economic structure and international competitiveness*, edited by John H. Dunning. Chichester: John Wiley and Sons.

Pae Chong-gi *et al.* (1990) *Chŏnja sanŏb-ŭi kyŏngjaengnyŏk-kwa kujo pyŏnhwa* (The competitiveness of and structural changes in the consumer electronics industry). Seoul: Sanŏp Yŏn'guwŏn.

Paek Chin-hyŏn. (1991) 'EC t'onghap-kwa taewoe kwan'gye-ŭi pyŏnhwa' (EC integration and changes in external relations), in *EC t'onghap-kwa uri-ŭi taeŭng pangan* (EC integration and our response), edited by Min Chung-ki. Seoul: Taewoe Kyŏngje Chŏngch'aek Yŏn'guwŏn.

Pak Ch'an-gwŏn. (1992) *Haewoe chikchŏp t'uja iron-jung naebuhwa iron-ŭi hwakchang moder-e kwanhan yŏn'gu* (A study of the extended model of the internalization theory in foreign direct investment theory). Master's Degree dissertation. Seoul: Myŏngji University Graduate School, Department of Economics.

Pak No-hong. (1994) *EC-e taehan ilbon, miguk, han'gug-ŭi haewoe chikchŏp t'uja chŏllyak pigyo yŏn'gu* (A comparative analysis of Japanese, US and Korean MNCs' strategies regarding the EC). Master's Degree dissertation. Seoul: Sŏnggyun'gwan University Graduate School of Trade, Department of International Economics.

Pak Sang-ho. (1996) *Algi shwiun haewoe t'uja shilmu* (Easy-to-understand overseas investment practice). Seoul: 21 Segi Buksŭ.

——. (1995) *Haewoe t'uja shilmu maenyuŏl* (An overseas investment practice manual). Seoul: 21 Segi Buksŭ.

Pak Se-un and Kim Ch'ang-sŏn. (1996) *Haewoe t'uja, woegugin t'uja* (Overseas investment, foreign inward investment). Seoul: Kukche Kŭmnyung Yŏn'guwŏn.

Pak Sŏng-t'aek *et al.* (1994) *Kajŏn sanŏb-ŭi haewoe chikchŏp t'uja chŏllyak* (The foreign direct investment strategy of the consumer electronics industry). Seoul: Sanŏp Yŏn'guwŏn.

Pak Yong-hwan. (1992) *Han'guk kajŏn kiŏb-ŭi kukchehwa chŏllyag-e kwanhan yŏn'gu* (A study of the internationalization strategy of the Korean consumer electronics industry). Master's Degree dissertation. Seoul: Chungang University Graduate School. Department of Foreign Trade.

Pan Pyŏng-gil. (1996) *Tagukchŏk kiŏp kwalli* (Management of the multinational enterprise). Seoul: Pagyŏngsa.

——. (1985) *Tagukchŏk kiŏmnon* (The theory of the multinational enterprise). Seoul: Pagyŏngsa.

Park Gi-han. (1996) *Foreign manufacturing investment in the European Union: The case of the Korean consumer electronics industry*. Doctoral thesis. Manchester Business School, University of Manchester.

Park Gi-han. (1995) 'Market entry strategy of Korean industry in the unified Euro-market', in *Industrial policy in Korea and the EU*, edited by Park Tae-kyu and Roy K. Wilkinson. Seoul: Institute of East and West Studies, Yonsei University.

Park Joon-Kyung. (1995) 'Creating extrafirm infrastructure of institutions for small and medium-sized businesses', in *Korea's choices in emerging global competition and cooperation*, edited by Cho Lee-jay and Kim Yoon Hyung. Seoul: Korea Development Institute.

Park Tae-kyu and Wilkinson, Roy K. (1995) *Industrial policy in Korea and the EU*. Seoul: Institute of East and West Studies, Yonsei University.

Park Tong-Whan. (Ed.) (1998) *The US and the two Koreas: A new triangle*. Boulder, Colorado: Lynne Rienner Publishers Inc.

——. (1998a) 'The Washington-Seoul-Pyongyang triangle and the future of the Korean peninsula', in *The US and the two Koreas: A new triangle*, edited by Park Tong Whan. Boulder, Colorado: Lynne Rienner Publishers Inc.

——. (1998b) 'Implications of Pyongyang's political-economic change', in *The US and the two Koreas: A new triangle*, edited by Park Tong Whan. Boulder, Colorado: Lynne Rienner Publishers Inc.

Park Young-ho. (1996) 'North Korean situation and prospects for inter-Korean relations', *Korea and World Affairs*, Spring, Vol. 20, No. 1, pp. 66–79.

Park Woo-sung. (2000) 'Korea's economy before and after the financial crisis', *Korea Focus*, Jan-Feb, Vol. 8, No. 1, pp. 60–70.

Pecht, Michael *et al.* (1997) *The Korean electronics industry*. New York, Boca Raton: CRC Press.

Penrose, Edith Tilton. (1956) 'Foreign investment and the growth of the firm', *Economic Journal*, Vol. 66. Reprinted in Dunning, John H. (1972) *International investment: Selected readings*. Harmondsworth: Penguin Modern Economics Readings.

Perrin, Serge. (1999) 'The emergence of Korean multinational enterprises in European manufacturing', in *The global integration of Europe and East Asia*, edited by Lee Sang-Gon and Pierre-Bruno Ruffini. Cheltenham: Edward Elgar.

Plunk, Daryl M. (1992) 'The continuing Cold War in Korea and US policy towards the peninsula in the 1990s', in *The US-South Korean Alliance: Time for a change*, edited by Doug Bandow and Ted Galen Carpenter. New Brunswick: Transaction Publishers.

Powlick, Philip J. (1998) 'US public opinion of the two Koreas', in *The US and the two Koreas: A new triangle*, edited by Park Tong Whan. Boulder, Colorado: Lynne Rienner Publishers Inc.

Pyŏk Yong-gi. (1993) *Tagukchŏk kiŏp kyŏngnyŏngnon* (Theory of multinational enterprise management). Seoul: Ilsa.

Randerson, Claire and Dent, Christopher M. (1996) 'Korean and Japanese foreign direct investment in Europe: An examination of comparable and contrasting patterns', *Asian Studies Review*, Vol. 20, No. 2, pp. 45–69.

Rees, David. (1988) *A short history of modern Korea*. Isle of Man: Ham Publishing Co. Ltd.

Rhee Sang-woo. (1990) 'Inter-Korea détente and the desirable role for the United States', in *Korea-US relations in a changing world*, edited by Robert Sutter and Han Sungjoo. Berkeley: University of California, Institute of East Asian Studies.

Rugman, Alan M. (1986) 'New theories of the MNE: An assessment of the internalization theory', *Bulletin of Economic Research*, 38, pp. 101–118.

——. (1980) 'Internalization as a general theory of FDI: A reappraisal of the literature', *Weltwirtschaftliches Archiv*, 116 (2), pp. 365–379.

——. (1975) 'Motives for foreign investment: the market imperfections and risk diversification hypothesis', *Journal of World Trade Law*, 9, pp. 567–573.

Rugman, Alan M. *et al.* (1985) *International business: Firm and environment*. Singapore: McGraw-Hill Book Company.

SaKong Il. (1999) 'Korea drafts vision of Europe-Asia relations', *Business Korea*, May, p. 55.

——. (1993) *Korea in the world economy*. Washington: Institute for International Economics.

Sanŏp Yŏn'guwŏn. (1990) *Kajŏn sanŏb-ŭi kukche kyŏngjaengnyŏk hwaktae-rŭl wihan yŏn'gu* (A study for the expansion of the international competitiveness of the consumer electronics industry). Seoul: Sanŏp Yŏn'guwŏn.

Scalapino, Robert A. (1986) 'Polemics and realities: US-North Korean relations', in *North Korea in a regional and global context*, edited by Scalapino, Robert A. and Lee Hongkoo. Berkeley: University of California, Institute of East Asian Studies.

Scalapino, Robert A. and Lee Hongkoo. (Ed.) (1986) *North Korea in a regional and global context*. Berkeley: University of California, Institute of East Asian Studies.

Shim In-sŏp. (1988) *Han'guk kajŏn sanŏb-ŭi kukchehwa kwajŏng-e kwanhan yŏn'gu* (A study of the internationalization process of the Korean consumer electronics industry). Master's Degree dissertation. Seoul: Yonsei University Graduate School of Business Administration.

Shin Jang-sup. (1998) 'Korea cannot rely solely on foreign investment', *Korea Focus*, Jul-Aug, Vol. 6, No. 4, pp. 16–34.

Shin Ki-t'aek. (1990) *Kyŏngje hwan'gyŏng-ŭi pyŏnhwa-wa hangug-ŭi haewoe t'uja chŏngch'aek panghyang* (Changes in the economic environment and the direction of Korea's overseas investment policy). Master's Degree dissertation. Seoul: Seoul National University Graduate School of Public Administration.

Shin Sang-hyup. (1998) *European integration and foreign direct investment in the EU: The case of the Korean consumer electronics industry*. London: Routledge.
Shin Sanŏp Kyŏngnyŏngwŏn. (Ed.) (1998) *30 taechaebŏl chaemu punsŏk* (An analysis of the financial status of the top 30 *chaebŏl*). Seoul: Shin Sanŏp Kyŏngnyŏngwŏn.
Shin Woong Shik and Oh, Eugene J. (1990). 'Recent developments in Korea's foreign investment', *Journal of World Trade*, Vol. 24, No. 6, pp. 31–56.
Shin Yong-dae and Yi Kyŏng-suk. (1993) *EC chinch'ul han'guk kiŏb-ŭi kyŏngnyŏng sŏnggwa punsŏk-kwa hyanghu t'uja chŏn'gae panghyang* (An analysis of the management results of Korean companies in the EC and future directions for investment). Seoul: Sanŏp Yŏn'guwŏn.
Slater, Jim and Strange, Roger. (Ed.) (1997) *Business relationships with East Asia: The European experience.* London: Routledge.
Snyder, Scott. (1997) 'Beyond the Geneva-agreed framework: A road map for normalizing relations with North Korea', in *Peace and security in Northeast Asia*, edited by Kihl Young Whan and Peter Hayes. New York: M. E. Sharpe.
Soh Byung-hee. (1997) 'Chaebol and politics: Past ills and future tasks', *Korea Focus*, May-June, pp. 56–65.
Soh Changrok. (1997) *From investment to innovation? The Korean political economy and changes in industrial competitiveness*. Seoul: Korea University Global Research Institute.
Sohn, Cecile. (1999) 'Analysts fear the W-shaped curve', *Business Korea*, June, pp. 16–17.
Sohn Jie-ae. (1994) 'Korea's big rush abroad', *Business Korea*, December, pp. 30–32.
——. (1993) 'Diversity or discord?', *Business Korea*, July, pp. 26–30.
Sohn Young-ju. (1999a) 'External factors threaten economy', *Business Korea*, April, pp. 19–20.
——. (1999b) 'Seven factors threaten trade surplus', *Business Korea*, January 1999, pp. 12–13.
——. (1997) 'Restructuring: Will it save Korea's falling dominoes?', *Business Korea*, November, pp. 22–23.
——. (1996a) 'Korean government's efforts for globalization in vain?', *Business Korea*, June, pp 21–22.
——. (1996b) 'Stable economic growth projected for 1996 at 7% level', *Business Korea*, May, pp. 18–19.
Son Sang-ik and Yi Kyŏng-suk. (1991) *EC t'onghap-e ttarŭn kiŏp mit sanŏp hwan'gyŏng-ŭi pyŏnhwa* (Changes in the corporate and industrial environment deriving from EC integration). Seoul: Sanŏp Yŏn'guwŏn.
Song Byung-nak. (1994) *The rise of the Korean economy*. Oxford: Oxford University Press.
Sŏnu Sŏk-ho *et al.* (1990) *Kajŏn sanŏb-ŭi kukche kyŏngjaengnyŏk chŭngdae-rŭl wihan yŏn'gu* (Enhancing the competitiveness of the consumer electronics industry). Seoul: Sanŏp Yŏn'guwŏn.
Spector, Leonard S. (1997) 'US-DPRK agreed framework on nuclear and related issues: Congressional testimony', in *Peace and security in Northeast Asia*, edited by Kihl Young Whan and Peter Hayes. New York: M. E. Sharpe.
Spicer, Andi. (1991) 'Scrambling to get through the door', *Korea Business World*, November, pp. 24–26.
Steers, Richard M. *et al.* (1989) *The Chaebol: Korea's new industrial might*. New York: Harper & Row.
Stokhof, Wim and Van der Velde, Paul. (Ed.) (1999) *ASEM: A window of opportunity*. London: Kegan Paul International.

Suarez-Villa, Luis and Han Pyo-hwan. (1990) 'The rise of Korea's electronics industry: technological change, growth and territorial distribution', *Economic Geography*, Vol. 66, pp. 273–292.

Suh Jang-won. (Ed.) (1993) *Korea's new international economic diplomacy and globalization policy*. Seoul: Korea Institute for International Economic Policy.

Suh Sang-Mok. (1998) 'The Korean currency crisis: What can we learn from it?', *Korea Journal*, Summer, Vol. 38, No. 2, pp. 34–55.

Sutter, Robert and Han Sungjoo. (1990) *Korea-US relations in a changing world*. Berkeley: University of California, Institute of East Asian Studies.

Tan, Joyce and Pelkmans, Jacques. (1991) 'Bilateral trade relations between Korea and the EC', in *The economic cooperation between EC and Korea: Problems and prospects*, edited by Min Chung-ki. Seoul: KIEP & Friedrich- Ebert-Stiftung Seminar, 14–15 October 1991.

Teece, David J. (1983) 'Technological and organisational factors in the theory of the multinational enterprise', in *The growth of international business,* edited by Mark Casson. London: George Allen and Unwin.

Tolentino, Paz Estrella E. (1993) *Technological innovation and third world multinationals*. London: Routledge.

U Sŏk-ho. (1985) *Tagukchŏk kiŏb-ŭi haewoe chikchŏp t'uja-e kwanhan yŏn'gu: Han, mi, ir-ŭi chungshim-ŭro* (A study of foreign direct investment by multinational enterprises: The cases of Korea, Japan and the United States). Master's Degree dissertation. Seoul: Chungang University, International Management School.

Ungson G.R. *et al.* (1997) *Korean enterprise: The quest for globalization*. Boston: Harvard Business School Press.

United Nations Conference on Trade and Development. (1997) *Sharing Asia's dynamism: Asian direct investment in the European Union*. New York: United Nations.

——. (1996) 'World investment report 1996: Investment, trade and international policy arrangements – Overview', *Transnational Corporations*, Vol. 5, No. 3, pp. 111–145.

United Nations Centre on Transnational Corporations. (1991) *World investment report 1991: The triad in foreign direct investment*. New York: United Nations.

Ursacki, Terry and Vertinsky, Ilan. (1994) 'Long-term changes in Korea's international trade and investment', *Pacific Affairs,* Vol. 67, pp. 385–409.

Vernon, Raymond. (1993) 'Where are the multinationals headed?' in *Foreign direct investment*, edited by Kenneth A. Froot. London: University of Chicago Press (National Bureau of Economic Research Project Report).

——. (1991) 'Sovereignty at bay: Twenty years after', *Millennium*, 20, pp. 191–196.

——. (1966) 'International investment and international trade in the product cycle', *Quarterly Journal of Economics,* Vol. 80. Reprinted in *The international political economy of direct foreign investment*, edited by Benjamin Gomes-Casseres and David B. Yoffie (1992), Aldershot: Edward Elgar Publishing Limited.

Waitt, Gordon. (1993) 'The composition and direction of the Republic of Korea's outward foreign direct investment', *Asian Profile,* Vol. 21, No. 3, pp. 201–215.

Wallace, Cynthia Day. (Ed.) (1988) *Foreign direct investment and the multinational enterprise: A bibliography*. Dordrecht, Boston, Lancaster: Martinus Nijhoff Publishers.

Wang Yun-jong (Ed.) (1997) *Han'gug-ŭi haewoe chikchŏp t'uja hyŏnghwang- gwa sŏnggwa: Shimch'ŭng pogo* (The current status and results of Korean foreign direct investment: An in-depth report). Seoul: Taewoe Kyŏngje Chŏngch'aek Yŏn'guwŏn.

Wells, Louis T. Jr. (1993) 'Mobile exporters: New foreign investors in Asia', in *Foreign direct investment*, edited by Kenneth A. Froot. London: University of Chicago Press (National Bureau of Economic Research Project Report).

——. (1983) *Third world multinationals: The rise of foreign investment from developing countries*. Boston: MIT Press.

——. (1981) 'Foreign investors from the third world', in *Multinationals from developing countries*, edited by Krishna Kumar and Maxwell G. McLeod. Lexington, Mass. and Toronto: Lexington Books.

——. (1977) 'The internationalization of firms from developing countries', in *Multinationals from small countries*, edited by Tamir Agmom and Charles P. Kindleberger. Cambridge: MIT Press.

Westerlund, Percy. (1999) 'Strengthening Euro-Asian relations: ASEM as a catalyst', in *ASEM: A window of opportunity*, edited by Wim Stokhof and Paul Van der Velde. London: Kegan Paul International.

Wilkins, Mira. (Ed) (1991) *The growth of multinationals*. Aldershot: Edward Elgar.

Williamson, Oliver E. (1981) 'The modern corporation: origins, evolution, attributes', *Journal of Economic Literature,* Vol. 19, No. 4, December, pp. 1537–1570

Windham-Wright, Thor. (1999) *The ASEM meetings and their effect on the relations between Japan and South Korea*. Undergraduate dissertation. Sheffield University: School of East Asian Studies.

Wŏn Chong-gŭn. (1987) 'Haewoe chikchŏp t'uja iron-ŭi munhŏnjŏk koch'al' (A bibliographical survey of foreign direct investment theory), *Muyŏk Hakhoeji*, Vol. 12, March, pp. 73–109.

——. (1986) *Han'guk kajŏn sanŏb-ŭi taemi chikchŏp t'uja-e kwanhan yŏn'gu* (A study of foreign direct investment in the United States by Korean consumer electronics companies). Doctoral thesis. Seoul: Koryo University Graduate School.

Yang Jun-sok. (2000) 'Korea's foreign trade', in *Korea's economy 2000,* edited by the Korea Economic Institute Editorial Board. Washington DC: Korea Economic Institute of America.

Yoo Cheong-mo. (1997) 'Financial stabilization steps unveiled', *Newsreview,* 13 December, pp. 12–13.

Yoo Jang-hee. (1993) 'A new internationalization strategy of the Korean economy: Suggestions and recommendations', in *Korea's new international economic diplomacy and globalization policy*, edited by Suh Jang-won. Seoul: Korea Institute for International Economic Policy.

Yoo S. J. (1999) 'Chaebols set massive downsizing plans', *Business Korea*, June, pp. 23–24.

Yoo Seong-min. (1998) 'Democracy, efficiency, equity and chaebol reform', *Korea Focus*, Jul-Aug, Vol. 6, No. 4, pp. 1–15.

Yoo Seong Min and Lee Sung Soon. (1997) 'Evolution of industrial organization and policy response in Korea: 1945–1995', in *The Korean economy 1945–1995: Performance and vision for the 21st century*, edited by Cha Dong-se *et al.* Seoul: Korea Development Institute.

Young Soogil and Kang Moonsoo. (Ed.) (1990) *The single European market and its implications for Korea as an NIE*. Seoul: Korea Development Institute & Friedrich-Ebert-Stiftung Conference, 18–19 June, 1990.

Yu C. J. and Ito K. (1988) 'Oligopolistic reaction and foreign direct investment', *Journal of International Business Studies,* 19 (3), pp. 449–460.

Yu In hak. (1991) *Han'guk chaebŏr-ŭi haebu* (An analysis of the Korean *chaebŏl*). Seoul: P'ulbit.

Yu Jae-chul. (1998) 'Where is the government's economic policy heading?' *Korea Focus*, May-Jun, Vol. 6, No. 3, pp. 17.

Yu Kwan-sŏn. (1994) *Han'guk kajŏn kiŏb-ŭi tae-EU chikchŏp t'uja pangan-e kwanhan yŏn'gu: D chŏnjasa-ŭi tae-p'urangsŭ t'uja sarye-rŭl chungshim-ŭro* (A study of direct investment by a Korean home appliance company in the EU: A case study of the investment in France by D. Electronics). Master's Degree dissertation. Seoul: Sogang University Graduate Business School.

Yu Man-sŏn. (1989) *Han'guk kajŏn'giŏb-ŭi taesŏnjin'guk chikchŏp t'uja chŏllyak panghyang* (The direction of the foreign direct investment strategies for developed countries adopted by Korean consumer electronics companies). Master's Degree dissertation. Seoul: Yonsei University Graduate Business School.

Yu Sunghoon. (1990) *Multinationals from Newly Industrializing Economies: The macro and microdeterminants of foreign direct investment by Korean manufacturing industries*. Doctoral thesis. George Washington University.

Yun Tong-jin. (1993) *Han'guk kajŏn kiŏb-ŭi kukchehwa kwajŏng-gwa kyŏngjaeng uwi* (The internationalization process and competitive advantage of Korean consumer electronics firms). Doctoral thesis. Seoul: Koryo University Graduate School, Department of International Business and Trade.

——. (1986) *Haewoe chikchŏp t'uja iron-ŭi t'onghaphwa-e taehan yŏn'gu: Munhŏnjŏk koch'ar-ŭl chungshim-ŭro* (A study of the integration of FDI theory: A bibliographical survey). Master's Degree dissertation. Seoul: Koryo University Graduate School, Department of Foreign Trade.

STATISTICS

Bank of Korea. (1997) *Haewoe t'uja t'onggye yŏnbo* (Overseas investment statistics yearbook.) Seoul: Bank of Korea.

Bank of Korea. (1996) *Haewoe t'uja t'onggye yŏnbo* (Overseas investment statistics yearbook.) Seoul: Bank of Korea.

Electronics Industries Association of Korea.(1997d) *Chŏnja chŏn'gi kongŏp kyŏnggi tonghyang* (Trends in the electronics/electrical industry). Seoul: EIAK.

——. (1997e) *Chŏnja kongŏp t'onggye* (Statistics of the electronics industry). Seoul: EIAK.

——. (1997f) *Chŏnja sanŏp haewoe t'uja ŏpch'e tirekt'ori* (A directory of overseas investors in the electronics industry). Seoul: EIAK.

——. (1997g) *Han'gugŭi chŏnja sanŏp* (The Korean electronics industry). Seoul: EIAK.

——. (1993a) *Chŏnja kongŏp t'onggye* (Statistics of the electronics industry). Seoul: EIAK.

——. (1993b) *1993 Han'gugŭi chŏnja kongŏp* (The Korean electronics manufacturing industry in 1993). Seoul: EIAK.

Federation of Korean Banks. (1998) *Haewoe chikchŏp t'uja t'onggye yŏnbo* (Foreign direct investment statistics yearbook). Seoul: Chŏn'guk Unhaeng Yŏnhap'oe.

Korea Foreign Trade Association. (1994) *Han'guk kyŏngje-ŭi chuyo chip'yo* (Key indices of the Korean economy). Seoul: Korea Foreign Trade Association.

Korea Overseas Information Service. (1993) *Facts about Korea*. Seoul: Korea Overseas Information Service.

National Statistics Office. (1997) *Han'guk chuyo kyŏngje chip'yo* (Major indices of the Korean economy). Seoul: National Statistics Office.

MAGAZINE ARTICLES

Baker & McKenzie European Union Newsletter

'Details of the EU-Korea Framework Agreement for Trade and Cooperation', May 1996.

Business Korea

'Hyundai family retires from management', June 2000, pp. 10–11.
'Man who never quits says he's quit', June 2000, pp. 12–13.
'Slow pace of reform threatens recovery', June 2000, pp. 18–20.
'No let up in reform drive', June 2000, pp. 22–25.
'Economy runs squarely on recovery track', May 2000, pp. 22–25.
'Manufacturers invest in expansion, improved productivity', May 2000, pp. 58–59.
'Government backs itself into a fiscal corner', April 2000, pp. 14–15.
'High oil prices and weak yen menace recovery prospects', March 2000, pp. 12–13.
'Move towards market mechanics is gradual', March 2000, pp. 14–15.
'President focuses economic leadership(sic)', February 2000, p. 8.
'New economic leader places stability first', February 2000, p. 9.
'Restructuring of domestic financial industry', February 2000, pp. 30–36
'Financial reform in Korea', February 2000, pp. 35–41.
'Foreign takeover may be mixed blessing', February 2000, pp. 64–66.
'Kia Motors: Back on the road', February 2000, pp. 66–67.
'Korea's recovery curve likely to level off', January 2000, pp. 14–15.
'Balance of trade might ride a roller coaster', January 2000, pp. 50–52.
'Creditors walk a maze to reorganize an insolvent chaebol', November 1999, pp. 20–24.
'Daewoo founder steps down from FKI leadership', November 1999, pp. 28–29.
'Prognosis looks healthier next year', October 1999, pp. 16–17.
'President Kim's economic policies bring mixed blessings', October 1999, pp. 18–19.
'Reasons for cautious optimism persist', October 1999, pp. 22–24.
'Corporate Korea takes a whipping', October 1999, pp. 30–32.
'Exports picture brightens for Korea', September 1999, pp. 14–16.
'President's approach might work', September 1999, pp. 20–21.
'President, crisis tame the chaebols', September 1999, pp. 22–24.
'Trade balance vulnerable to many factors', July 1999, pp. 12–14.
'Government shifts towards not rocking the boat', July 1999, pp. 15–16.
'Improved rating reflects growing confidence in Korea', July 1999, p. 74.
'Chaebol suddenly reverses course', May 1999, p. 19.
'Korea's top chaebol will downsize radically', May 1999, p. 20.
'Financial reform takes aim at second tier', April 1999, pp. 17–18.
'Korea's export potential erodes', April 1999, p. 22.
'Chaebols move to expand empires', April 1999, p. 43.
'Nation collects financial vote of confidence', March 1999, pp. 10–11.
'Long-term recovery is goal', March 1999, pp. 23–24.
'Korea rolls out red carpet to foreign investors', March 1999, p. 60.
'Seeds are sown for economic recovery', January 1999, p. 16.
'Early downsizing means survival', November 1998
'No danger of second crisis, President says', October 1998, p. 12.
'Several hurdles remain in banking reform', October 1998, pp. 28–29.

'Causes and solutions to the economic crisis', October 1998, pp. 54–58.
'IMF Report: Update on Korea's reform', October 1998, pp. 60–61.
'Government seeks ways to lure sagging direct investment', October 1998, p. 63.
'Major economic indicators', October 1998, p. 82.
'Maybe right leader at right time', September 1998, pp. 42–43.
'Government, Big Five fight over reform', August 1998, p. 12.
'10 core industries targeted for shakedown', August 1998, pp. 12–13.
'Korea, IMF agree to ease recession', August 1998, p. 20.
'Reformed economy can restore growth in 1999', August 1998, p. 22.
'Second half looks even worse', July 1998, p. 13.
'Crumbling markets', July 1998, pp. 14–15.
'No will for reform', July 1998, pp. 17–18.
'Direct investment picking up', July 1998, pp. 42–43.
'The toughest reforms are still ahead', July 1998, p. 54.
'Dismantling another conglomerate', June 1998, pp. 13–14.
'A big bang is imminent', June 1998, p. 65.
'Violent labour strife threatens recovery', May 1998, p. 11.
'Heavy borrowing backfires', May 1998, p. 25.
'Foreign direct investment liberalization plan', May 1998, p. 34.
'A Korean-style market economy?', April 1998, p. 13.
'Steady as she goes', April 1998, p. 17.
'Small is beautiful', April 1998, pp. 44–45.
'High interest rates and massive unemployment', March 1998, p. 13.
'Korea's reforms in the spotlight', March 1998, p. 39.
'President Kim's inaugural address: "Government of the people"', March 1998, pp. 40–42.
'The chaebol have no choice', February 1998, p. 21.
'Ten insolvent players shut down', February 1998, p. 52.
'Japanese products to flood Korean market', January 1998, p. 12.
'Korean banks struggling with problem loans', January 1998, pp. 12–13.
'A tough beginning', January 1998, pp. 18–19.
'Daewoo takes over Ssangyong Motor', January 1998, pp. 21–22.
'The market is now wide open', January 1998, pp. 35–36.
'Slower growth, higher unemployment', January 1998, pp. 36–37.
'Privatization efforts stalled', January 1998, p. 46.
'String of mergers and acquisitions expected', January 1998, p. 53.
'Making the grade', January 1998, p. 55.
'Do or die', December 1997, pp. 22–23.
'Dealing with the IMF', December 1997, pp. 27–28.
'Halla goes under', December 1997, pp. 28–29.
'Troubled, but basically sound', December 1997, pp. 29–31.
'Modest recovery ahead', November 1997, p. 19.
'1998 will definitely be better', November 1997, pp. 20–21.
'Slow growth forecast', October 1997, p. 13.
'Troubled investments', October 1997, pp. 17–18.
'Electronics industry policy: Diversify, diversify', October 1997, pp. 35–37.
'Showing smart electronic products', October 1997, p. 38.
'Country report: The UK – Bicentennial celebration', October 1997, pp. 68–74.
'Borrowing difficulties', September 1997, pp. 24–25.
'Steadying the rocky financial market', September 1997, p. 25.
'Survival of the fittest', September 1997, p. 38.
'Kia Motors: What a mess!', September 1997, pp. 40–41.

'Saddled with debt', June 1997, p. 12.
'Slower recovery, slower growth (Asian Development Bank report on Korea)', May 1997, pp. 15–17.
'An interview with the Deputy Prime Minister: Tightening the government belt', May 1997, pp. 22–24.
'A new economic team has a tough assignment', April 1997, p. 13.
'New policy directives for the troubled economy', April 1997, pp. 16–17.
'Economic policy must change', March 1997, p. 17.
'Hanbo goes under', February 1997, pp. 12–13.
'Opening up: The government deregulates the Korean economy', October 1996, pp. 14–17.
'Alarming trade deficit control', April 1996, p. 17.
'Korean exports to the UK rise 61%, British exports 44%', April 1996, p. 54.
'ASEM: Asia, Europe lay foundation for mutual understanding and cooperation', April 1996, pp. 56–58.
'Outgoing to meet new flooding challenges', March 1996, pp. 17–21.
'Korea's directions and policies for the economy in 1996', February 1996, pp. 16–19.
'Economic outlook for 1996', January 1996, pp. 16–17.
'Top stories in 1995 & looking ahead to 1996', January 1996, pp. 18–21.
'Korea's industries in 2000s', January 1996, pp. 36–38.
'Trade and industrial policies for the new economy', October 1995, pp. 13–15.
'This year's momentum to carry over into 1996: Slowdown, but soft landing projected', October 1995, pp. 16–17.
'50 years of independence and division: Looking back and ahead', September 1995, pp. 10–11.
'Exports to break US$100 billion mark', February 1995, pp. 17–18.
'Government struggles to hold down inflation', September 1994, pp. 37–38.
'Textiles and electronics to recover', July 1994, pp. 26–27.
'Korean economy in expansion phase: BOK', April 1994, pp. 31–32.
'Paper tiger?', April 1993, pp. 16–17.
'Seeking to regain growth potential', April 1993, pp. 14–15.
'Less government controls for an economic revival', March 1993, pp. 16–18.
'Gradual recovery', January 1993, pp. 30–31.
'1993 business outlook: Electronics', January 1993, p. 47.

Economic Report

'Government will likely revise upwards economic growth forecast to 5%', July 1999, pp. 21–22.
'Chaebol post 68.9% hike in investment', July 1999, pp. 40–41.
'Conglomerates to slash investments in 1998', December 1997, p. 49.
'EU calls on Korea to speed up financial service liberalization', November 1997, p. 55.
'Economy projected to grow by 6.5–7 pct. next year', October 1997, pp. 34–35.
'BOK forecasts narrower current account deficit in 1997', September 1997, p. 32.
'June exports hit record high', August 1997, p. 23.
'Kia Group selected in antibankruptcy arrangement', August 1997, p. 30.
'US, EU call for further opening of ROK car market', August 1997, p. 40.
'EU, Korea to settle disputes over chip exports', August 1997, p. 40.
'South Korea's economy expected to recover in second half', July 1997, pp. 22–23.
'Hanbo scandal shakes S. Korean banks to root', March 1997, pp. 16–18.

'Hanbo demise drives banks, SMEs into total mess', March 1997, p. 19.
'Rotten at the top; Lawmakers, bankers arrested over Hanbo scandal', March 1997, p. 14.
'6.5% for S. Korean economic growth this year', January 1997, pp. 16–17.
'Conglomerates push overseas operations, growing into MNCs', January 1997, pp. 26–27.
'South Korea records US$16.8 billion in trade deficit', December 1996, pp. 29–30.
'Pros and cons mark entry to advanced nations' club', November 1996, pp. 26–27.
'Daewoo Electronics acquires Thomson Multimedia', November 1996, p. 47.
'Wales attracts largest Korean investment ever', August 1996, pp. 16–17.
'1995 economy extremely healthy', May 1996, pp. 10–11.
'Chaebols expand world wide', May 1996, pp. 15–17.
'Current Korean trade balance troubling', April 1996, pp. 10–11.
'Diversification pays off', March 1996, pp. 44–45.
'Korea's direct investment at all-time high', December 1995, pp. 24–25.
'Korea's economy cooling off in 1996?', December 1995, pp. 28–29.
'Forecast good for East Asia economy', November 1995, pp. 45–46.
'Sunny days ahead for the global economy?', November 1995, pp. 33–35.
'Hermit Kingdom is history', February 1995, pp. 21–22.
'Great expectations', October 1994, p. 47.
'Act your age', October 1994, p. 46.
'Electronics: Growing sophistication', July 1994, p. 23.
'Kim's economic policy: Not good enough', March 1994, pp. 18–19.
'Consumer electronics: Mild growth', December 1993, p. 35.
'Electronics: A yen for exports', August 1993, p. 56.
'Reserving judgement', June 1993, p. 53.
'In the name of reform', May 1993, p. 25.

Economist Archives (http://www.economist.com/archive/)

'To the brink, and back again', 3 June 2000.
'Silver lining', 1 April 2000.
'Career path', 1 April 2000.
'Victims of success', 6 November 1999.
'Lost Seoul', 4 September 1999.
'The death of Daewoo', 21 August 1999.
'South Korea's alarming recovery', 31 July 1999.
'Capital flaws', 10 July 1999.
'Nationbuilders', 10 July 1999.
'Lazarus of the East', 29 May 1999.
'South Korea's workers return to the streets', 1 May 1999.
'More pain please', 17 April 1999.
'Policy failures', 20 March 1999.
'Seoul survivors', 30 January 1999.
'Thinking small', 14 November 1998.
'The chaebol that ate Korea', 14 November 1998.
'Boom? What boom?', 12 September 1998.
'Trouble upon trouble for Kim', 29 August 1998.
'Forced friendship in South Korea', 8 August 1998.
'Poor health', 11 July 1998.
'The fight ahead', 13 June 1998.
'Brave talk', 23 May 1998.

'Korean glums', 10 January 1998.
'Web site', 13 December 1997.
'A fatal resolve not to budge', 22 November 1997.
'Vicious circle', 25 October 1997.
'The giants stumble', 18 October 1997.
'Bad habits', 13 September 1997.
'The Kia standard', 6 September 1997.
'First and worst', 23 August 1997.
'Kia keels over', 19 July 1997.
'No bang', 7 June 1997.
'Lessons unlearnt', 21 December 1996.

Far Eastern Economic Review Archives (http://www.feer.com/)

'E-Asia: The year of the Internet', 30 December 1999.
'Korea's chance', 4 November 1999.
'Battle of wills', 26 August 1999.
'Cause and effect', 1 July 1999.
'Size doesn't matter', 4 March 1999.
'New kid in town', 11 February 1999.
'Problems unaddressed', 21 January 1999.

Korea Economic Weekly

'Economic relations between Korea and European Union expected to strengthen', No. 329, 13 February 1995, p. 3.

Korea Herald (http://koreaherald.co.kr/)

'Seoul unveils second-half economic management plan', 24 June 2000.
'Obstacles for Korean economy in 2000', 17 January 2000.

Korea Times

'End of Kim's European visit', 15 March 1995, p. 2.
'President Kim's Europe visit', 3 March 1995, p. 2.
'Correction of trade imbalance tops President Kim's agenda for Europe visit', 3 March 1995, p. 6.

Korea Trade Update

'Reinventing the Korean economy through reform and market opening', Issue 1, November 1999, pp. 2–4.

Newsreview

'Local electronics makers' offshore production up 32%', 18 April 1998, p. 23.
'Incoming administration working to enact reforms asked by IMF', 27 December 1997, pp. 10–11.
'IMF G-7 to loan US$10 billion in emergency aid to Seoul', 27 December 1997, pp. 12–13.
'Businesses on track to rebirth after bankruptcies', 27 December 1997, pp. 10–11.

'President-elect Kim D. J. stresses economic recovery, stability', 20 December 1997, pp. 6–7.
'Economic policies to change within IMF framework under new government', 20 December 1997, p. 7.
'Korea will be able to repay IMF loans by 2005 if it carries out agreement', 13 December 1997, p. 13.
'Three major parties support IMF deal', 6 December 1997. p. 7.
'Seoul, IMF agree on US$55 billion in bailout loans', 6 December 1997, pp. 12–15.
'Conglomerates pressured to launch massive restructuring steps', 6 December 1997, p. 13.
'IMF mission here for talks on bailout loans', 29 November 1997, p. 24.
'IMF bailout will bring about bankruptcies of marginal firms, drop in real estate prices', 29 November 1997, p. 25.
'IMF bailout hurts Korean pride', 29 November 1997, p. 26.
'Korea, EU agree on early negotiations for industrial pact', 8 November 1997, p. 10.
'Kia crisis', 26 July 1997, pp. 14–15.
'Accord with EU sets framework on cooperation', 9 March 1996, p. 4.

Republic of Korea Economic Bulletin

'Major financial policies and measures taken during January 2000', February 2000, pp. 25–26.
'Financial reform in Korea', February 2000, pp. 35–41.
'Policy directions for 2000: Reform update and macroeconomic stability', February 2000, pp. 42–44.
'Economic trends: Monthly review', December 1999, pp. 2–3.
'Final results of Korean government's second consultation with IMF for 1999', December 1999, pp. 23–45.
'Korea will continue to push economic reform and restructuring', December 1999, pp. 56–60.
'Additional measures for chaebol reform', September 1999, pp. 24–29.

WEBSITES

http://www.imf.org/external/np/sec/nb/1999/nb9985.htm
http://www.imf.org/external/np/speeches/1998/012298.htm
http://www.imf.org/external/np/speeches/1999/011699.htm

Index

Aharoni, Y. 8
An Ki-ch'ŏl 120, 122
An Se-yŏng 19 *see also* Competitive Advantage-Representative Specialization Zone Theory
Anam Industrial (company) 77
'Angel Clubs' 180
anti-dumping action *see* non-tariff trade barriers
anti-trade oriented investment 11
Asia Development Bank 176
Asia-Europe Meeting (ASEM) 112–14, 128
Asia-Pacific Economic Cooperation (APEC) agreement 112
Asian Games (1986) 65
Association of South-east Asian Nations (ASEAN) 187–88

'back door' entry to new markets 15, 142
Bank of Korea 51, 82, 91, 130, 177
banking system 177–78, 189
bankruptcies 48, 54, 177, 179, 182
'Big Deals', (action for corporate reform 1997–) 84, 180–81
'Big Three' consumer electronics companies 77–78, 154–59, 172 *see also* Daewoo Electronics, Goldstar, LG Electronics, Samsung Electronics
brand names, development of 74–75, 165, 173
Bridges, Brian 113
Buckley, Peter J. 6, 8–9, 171
Bulgaria
  diplomatic ties with Korea 126
business environment
  1963–1986 160–63
  1987–1996 163–70
  1997– 187–92
Business Week 181
Byun Jae-woong 32, 58, 148

capability-seeking investment 11–12
Casson, Mark 8–9, 171
central Europe
  Korea's relations with 94, 129, 168, 174
  Korean investment in 99, 105, 131–35, 159, 172–75
chaebŏl (conglomerates) 49–56, 76–77, 178–83
  bankruptcies in 1997 54, 179
  criticism of 52–53, 166
  definition of 49
  factors in development and growth 50–52
  FDI activity by 83–4, 87, 93, 105, 130
  government policy towards 51–55, 160–61, 167, 170, 179–82
  management style 179, 181
  reform and restructuring of 55, 178–83, 192
  relationship with government 51–54
Chang Myŏn 39
Chi Nam-ung 63, 147–48
China, People's Republic of
  Korea's relations with 46, 107–8
  Korean investment in 24, 99, 145

Korean trade with 47
'*chiptchung houshik*' ('flooding the market') *see* export strategy
Cho Soon 38
Cho Wŏn-dong 110, 117
Choi Yong-rok 23–24
Chŏng Sŏng-ch'ŏl 121
Chun Doo-hwan 43, 52, 73, 109–10
Chung Ju-yung 182
Ch'a Wŏn-dae 58
Ch'ae Chŏm-gil 22
Commercial Bank of Korea 177
competitive advantage
erosion of Korea's 66–68, 79
in FDI 6, 31
of Korean exporters 94
of Korean investors 167
Competitive Advantage-Representative Specialization Zone Theory (An Se-yŏng) 25, 30–33
components industries *see* parts and components industries
'Comprehensive Measures for Economic Stabilization' 43
Confucian social system 161
conglomerates *see chaebŏl*
consumer electronics industry 32–33, 57–81, 152–55, 169
characteristics of 71–81, 152–53, 169–70, 172–74
export dependency 72
FDI by 142–47, 160
foreign participation in 62, 72–75, 152, 170
government support for 63–65, 70–71, 169
import dependency 70, 72
oligopolistic structure 77–78
phases of development 61–66, 69–71
place in global industry 60
technological level of 75–76
see also OEM sales, parts and components industries
Corporate Restructuring Control Commission 180
'cost of foreignness' 5–6, 16, 26, 34, 152, 171
credit rating agencies 48, 189
cross-payment guarantees 179
Czech Republic
Korea's relations with 126–27
Korean trade with 128
Daewoo Electronics 1, 63, 77, 130, 142–43, 171–72
investment in Europe 154–59
proposed sale of 182–83
Dainong (company) 54
democratic reforms (1987) 45, 53, 66, 163, 173
deregulation 166–69, 186 *see also* liberalization
development-oriented investment 11
'Djnomics' 180
dumping *see* non-tariff barriers and trade friction
Dunning, John H. 8, 10–13, 16, 19–21, 171

eastern Europe
Korea's relations with 94, 125–29, 168, 174
Korean investment in 99, 105, 131–35, 159, 172–75
EC Communiqué on EC-Korean relations 110
Eclectic Paradigm of International Production (Dunning) 8, 10, 16, 20, 30–31, 171
Economic Cooperation Development Fund 127
economic crisis, Asian 3, 48, 83, 174
economic crisis, Korean (1997–) 48–49, 176–191
problems facing the economy 190–92
economic development
(1948–1997) 36–49
(1997–) 187–191
financing of development 39
economic development plans 38, 40, 47
economic policy 39, 43, 47, 160–61, 164, 166, 173
in aftermath of 1997 crisis 187–89
economies of scale (in production) 6–7, 29, 72, 79, 118, 161–63, 166
The Economist 181
education, expansion of 38
efficiency-seeking investment 12, 94, 105
Electronics Industries Association of Korea (EIAK) 2, 58, 75, 143, 145, 154
electronics industry
ranking in global exports 60
ranking in global production 59
structure of 59, 69

see also consumer electronics industry, industrial electronics
Electronics Industry, Long-Term Plan for Enhancement of 65, 169
Electronics Industry Promotion Fund 65
Electronics Industry Promotion Law 62, 169
ethnic minorities, Korean overseas 24, 142, 145
ethnic tie theory 24
EU Framework Agreement for Trade and Cooperation (with Korea) 111–13
Europe
  as counterbalance to the USA and Japan 109, 111, 119, 147, 168
  investment in Korea by 118–19
  Korea's relations with 107–29
  Korean investment in 130–50 *see also* foreign direct investment, Korean
  Korean trade with 114–18
  trade restrictions in 109, 116–18, 122
  see also Asia-Europe Meeting, central Europe and eastern Europe
European integration, Korean view of 119–26
European Union (EU) 70, 94, 105, 165–69, 174–75
  Common Commercial Policy 116
  expansion of membership 124–25
  Single European Market 119–21, 124, 128, 131, 141
export price competitiveness 79, 105, 152, 173
  erosion of 45, 66–68, 147, 163
export promotion measures 27–28, 32, 39–40, 63–64, 150, 152, 160–61, 173
export strategy 79, 169–70, 173

Fair Trade Act 52
Fair Trade Commission 54
Far Eastern Economic Review 184
financial sector *see* reforms, in 'IMF' era
Financial Supervisory Commission 177, 180
firm-specific advantages 6, 10, 23–24, 30–33
  of firms from developing countries 14–15, 22, 24
  of Korean firms 25–29, 76, 148, 151–53, 160, 171, 175
follow-the-leader patterns of foreign direct investment by Korean firms 33, 78, 151, 153, 156, 160, 170
Foreign Capital Inducement Act 61
foreign direct investment (FDI)
  by developing countries 12–15, 19–20, 24–25
  by industrialized countries 23
foreign direct investment in Korea 61–62, 73–75, 80–81, 185–186
foreign direct investment, Korean 82–106, 130–150
  by individuals 101
  by SMEs 87–88, 93, 99, 101–2, 105
  characteristics of 22, 28, 34, 151
  chronological overview 83–94
  equity participation ratio 95, 141
  government policy on 84–92, 104–5, 167, 173
  in developing countries 101–4
  in Europe 88, 93, 130–50
  in industrialized countries 101–4
  in manufacturing 96–101, 105, 134–40, 150
  joint ventures 95, 146, 149
  'pull' factors in 106, 174–75
  'push' factors in 106
  regional analysis of 85, 87–88, 92–93, 98–101, 131–32
  related to exports 104, 142
  scale of investment 94–95, 140
  sectoral analysis of 85, 86, 92, 96, 99–104, 134–40
  see also Theoretical/macroeconomic framework for Korean foreign direct investment
foreign direct investment, motivations for
  firms from developing countries 15
  firms from industrialized countries 11–12
  Korean firms 33, 94, 141–42, 147–50
Foreign Investment Promotion Act 186
'Fortress Europe' 119–24, 129, 131, 149, 160, 165
'four plus one' reforms 177

General System of Preferences (GSP) 116
General Trading Companies (GTCs) 42, 50–51, 90, 130, 150

globalization
of the Korean economy 47, 91, 105, 126, 129, 166, 168
role of FDI in 89, 94, 141–50, 176
Goh Chok Tong 112
Goldstar 61, 63–64, 77, 130, 143, 153–54 *see also* LG Electronics
Group of Seven (G7) nations 112

Ha Pyŏng-gi 101
Haitai Electronics 77
Han Chong-su 109
Hanbi Sanŏp (company) 130
Hanbo Iron and Steel 54
Hanil Bank 177
Havel, Vaclav 127
Heavy and Chemical Industry Plan (HCIP) 41–43, 56
Heenan, D. A. 4
Hitachi 61
Hoechst 118
'hollowing out' phenomenon in employment 88
Hong Sŏg-il 101
human rights 109–11
Hungary
Korea's relations with 126–27
Korean trade with 128
Hŭngmyŏng Kongŏp (company) 130
Hymer, Stephen 6, 31, 171
Hyundai Electronics 55, 77, 182–84

imperfect markets 5–6, 9, 26
import restrictions *see* trade restrictions
Import Source Diversification Programme 185
import substitution 29, 38, 62
indirect investment 5
Industrial Bank of Korea 53, 90
industrial electronics 60, 64, 65, 144
industrial organization theory 6, 8
industrial structure 28, 37, 88, 190
imbalances in 42
Integrated Theoretical Model (Yun Tong-jin) 25–28
intellectual property rights 118
intermediate products 9
internal markets of firms 2, 8–9
internalization incentive advantages 10
internalization theory 2, 8–10, 23–26, 30–31, 34, 154, 171–72
International Monetary Fund (IMF) 3, 54–55, 176, 181, 187–91
international relations 168–69 *see also* (central and eastern) Europe, North Korea, United States
international trade theory 5
Investment Development Cycle (Dunning) 12–14, 19, 21
'involuntary internationalization' 2–3, 34, 160, 172–74

Japan
colonial rule by 108
compensation from 39
Korean trade with 74, 78–79, 108–9, 185
Jeong Yeo-cheon 142
Jinro (company) 54
joint ventures *see* foreign direct investment, Korean
Jun Yongwook 2, 19, 22, 29, 32, 57–58, 152–54 *see also* Structural Market Defence Theory and 'structural irreversibility'

Keegan, W. J. 4
Kia Business Group 54
Kim Bak-soo 121
Kim Chin-yŏng 58
Kim Chong-tae 181, 185
Kim Dae-jung 110, 176, 180–81, 185–86, 189–90
Kim Ji-hong 123
Kim Kyu-ch'ang 33
Kim Sŏk-hŭi 58, 71, 73, 77
Kim Sŭng-jin 101–4
Kim Woo-choong 181–82
Kim Young-sam 46–47, 53–54, 111, 127, 166
Kindleberger, Charles P. 6
Knickerbocker, F. T. 2, 7–8, 16, 33, 153
Kojima, Kiyoshi 8, 10–11, 21
KOMI Semiconductors 62
Koo Bohn-young 19
Korea Asset Management Corporation 177
Korea Development Corporation 84
Korea Electric Power Company 183
Korea Exchange Bank 189
Korea First Bank 177
Korea Foreign Trade Association 114, 117–18

Korea Gas Corporation 183
Korea Herald 190
Korea Institute for Economics and Technology 58
Korea Investment Services Centre 186
Korea National (company) 63
Korea Telecom 183
Korean War 38, 50–51, 61, 108–9
Ku Cha-yong 58

labour costs
as competitive advantage 22, 39, 80, 152, 161, 171
as incentive for FDI 62, 140–42, 145, 147–48, 166, 174
increases in 34, 42, 45, 66–67, 145, 147, 163, 166–67, 173
labour disputes 45, 167, 184
labour market *see* reforms, in 'IMF' era
labour shortages 163, 173
land prices 163
land reform 38
land sales to foreigners 186
'leapfrog' patterns of foreign direct investment 159–60, 170
Lee Chae-min 82
Lee Sŏn-sang 58
LG Electronics 77, 171–72, 183
investment in Europe 153–59
see also Goldstar
liberalization
of financial sector 177–78
of inward investment 91, 166–67, 177, 185–86
of trade 166–67, 177, 185–86
lifetime employment 184
local content requirements *see* non-tariff trade barriers
localization theory 23–24, 147
location-specific advantages 10, 16, 23–24, 31, 172–75
Lomé Convention 116

manufacturing, investment in *see* foreign direct investment, Korean
market-seeking investment 11–12, 94, 105
marketing and distribution networks 74–75, 80, 105, 165, 173
mergers and acquisitions (M&A) 95, 105, 149, 186
Min Chung-ki 122
minimum price strategy theory 23–24
minor innovation theory 24
monetary policy 187–88
monopolistic advantage theory 30–31
motivations for investment *see* foreign direct investment, motivations for
multinational enterprises, from developing countries 14–16
multinational enterprises, theories of 4, 8–9

Nam Duck-woo 110
Nam Sang-kun 33, 126
net outward investment (NOI) 13–14
Newly Industrializing Economies (NIEs) 36, 120, 148, 165
Nixon Doctrine 41
non-tariff trade barriers 68, 109, 121–22, 165, 167
anti-dumping action 65, 68, 116–18, 122, 148
local content requirements 22, 116–17, 122
screwdriver assembly plants 116–17
North American Free Trade Agreement (NAFTA) 70, 94, 165
North Korea 37, 56, 108, 113, 161, 168–69
Reunification with South Korea 111, 114
'Northern Diplomacy' 99, 126

OEM (original equipment manufacture) sales 73–75, 152–53, 163, 165, 170, 173
oil shocks 41, 86
OLI (ownership-localization-internalization) Paradigm (Dunning) 171
Oligopolistic Reaction Theory (Knickerbocker) 2 , 8, 20, 32, 153, 157–60, 172
oligopoly
in FDI theory 7–8 *see also* Oligopolistic Reaction Theory
in Korean consumer electronics industry 33, 77–78, 81, 148, 151, 153, 170, 173, 175
Olympic Games, Seoul (1988) 65
Ŏm Ch'ang-sŏp 121, 124
ombudsman for foreign investors in Korea 186

'Opportunity Europe' 124–26
Organization for Economic Cooperation and Development (OECD) 36, 48, 91, 111, 116, 166, 183, 191–92
Overseas Investment Screening Council (OISC) 89–92
ownership-specific advantages 10
Ozawa Terutomo 21

Pae Chong-gi 67, 74
Paek Chin-hyŏn 123
Pak Sŏng-t'aek 58
Pak Tong-bae 121
Pak Yong-hwan 58
Pan Pyŏng-gil 19–20, 33
Park Chung-hee 39–41, 43, 50–51, 109, 161
Park Gi-han 32–33, 58, 153, 156, 172
parts and components industries 76–77, 144, 153, 170, 175
Pecking Order Theory (Hufbauer) 14, 16, 22, 158
Plaza Agreement (1985) 44
Pohang Iron and Steel 183
Poland
  Korea's relations with 126–27
  Korean trade with 128
political reforms *see* democratic reforms
political system environment theory 24–25
portfolio investment 5
Post-Product Life Cycle Model (Yun Tong-jin) 25, 28–30, 62–66, 70
price competitiveness *see* export price competitiveness
privatization 43, 142, 174, 181, 183
Product Life Cycle Model (Vernon) 6–7, 15, 28
  revised theory 7

real estate *see* land sales
reciprocity, in market access 123
reforms, in 'IMF era' 176–86
  corporate sector 179–83
  financial sector 177–78
  labour market 184
  public sector 183
regional economic blocs 70, 94, 121, 165, 167 *see also* EU and NAFTA
Regulatory Reform Committee 183
research and development (R&D) 120
  collaboration in 125, 127
  in consumer electronics industry 65, 67–68, 70, 74–76, 165, 170, 175
resource-seeking investment 11–12, 86, 88, 94, 105
reverse direct investment (RDI) 2, 18–34, 57, 158, 160, 172
Rhee Syngman 38–39, 50
Roh T'ae-woo 52–53, 110, 126, 167
royalty payments 76–77
Rugman, Alan M. 8

Sammi Steel 54
Samsung Electronics 63, 77, 130, 143, 171–72, 183
  investment in Europe 153–55, 158–59
screwdriver assembly plants *see* non-tariff trade barriers
segyehwa see globalization
Seoul Bank 177
Shim In-sŏp 58
Shin Jang-sup 148
Shin Ki-t'aek 22–23
Shin Sang-hyup 57–58
Slater, Jim 33, 126
small- and medium-sized enterprises (SMEs) 76–77, 141, 179–80
  government support for 53, 166, 180, 188
  restrictions on development 52–53
  see also foreign direct investment, Korean
Sŏnu Sŏk-ho 68, 147
Sony Corporation 61
strategic capability-seeking investment 11–12, 94, 105
'structural irreversibility' 2, 26–27, 29, 34, 154, 160, 172
Structural Market Defence Theory (Jun Yongwook) 2, 25–28, 172
such'ul ipkuk ('nation building through exports') 39 *see also* economic policy
sunset industries
  in Europe 93, 120

technical manpower, lack of 68
technical standards, harmonization of in EU 122–23, 125
technological gap model 14

technology
as advantage in FDI 14–15, 22, 24, 152, 175
as incentive for FDI 113, 129, 142, 145–46, 149–50, 167–68, 170, 173
in consumer electronics industry 75–76, 80–81, 152–53, 170, 175
technology-seeking investment 31
Theoretical/macroeconomic framework for Korean foreign direct investment in Europe 2–3, 34–35, 56, 150–51, 154, 160–70, 175
theories of foreign direct investment (FDI) 4–17, 25–33, 151–52, 171
early theories of foreign production 4
Japanese theories 10–11
Korean critiques of western and Japanese theories 18–33, 151, 171
Korean theories 25–33, 151, 171–72
theories of developing country FDI 12–15, 24–25
Western theories 4–10
see also theories of the multinational enterprise
theories of the multinational enterprise 5, 8–11
theory of the firm 8
'Three Lows' ('Three Blessings') 44, 163
Tokyo Round 68
trade creation and diversion in Single European Market 121
trade friction 44, 64–65, 67–68, 79–81, 88, 94, 145, 163, 165, 167, 173, 185
as incentive for FDI 26–27, 31–34, 147, 150, 173
with Europe 65, 79–81, 88, 109, 116–18, 131, 145, 148–49
trade, liberalization of *see* reforms, in 'IMF' era
trade-oriented investment 11
trade restrictions *see* non-tariff barriers and trade friction
trade unions 45, 66, 163, 173, 190
transaction cost economics 8–9, 146
Tripartite Commission (government-management-labour) 184

unemployment 184, 187–90
Ungson, G. R. 142
United Nations 50, 109
Conference on Trade and Development (UNCTAD) 130, 142, 145–46, 165
Security Council 111
United States
aid to Korea 38, 50, 108
investment in Korea 5–6, 11
Korean reverse direct investment in 20–21, 24, 26
policy towards Asia 41, 113
policy towards North Korea 113–14, 168–9
trade with Korea 44, 63–64, 74, 78–79, 108–9
Uruguay Round 165

Vernon, Raymond 6–7, 15, 28, 31, 175
Vietnam War 39, 41, 51, 86
Voluntary Export Restraints 64, 68

wages *see* labour costs
Waitt, Gordon 89
Walesa, Lech 127
Wells, Louis T. 14–15, 22
Williamson, Oliver E. 9
Wŏn Chong-gŭn 20–22, 147, 153, 172
theoretical explanation of FDI 21
work ethic, Korean 161, 163, 173
World Bank 176
World Trade Organization (WTO) 56, 112, 114, 165, 167, 183

Yu Kwan-sŏn 58
Yun Tong-jin 2, 19, 34, 152, 154, 160, 172 *see also* 'involuntary internationalization', Integrated Theoretical Model and Post-Product Life Cycle Model
Yushin (Revitalizing) Reforms 41